Memristors - The Fourth Fundamental Circuit Element - Theory, Device, and Applications

Edited by Yao-Feng Chang

Published in London, United Kingdom

Memristors - The Fourth Fundamental Circuit Element - Theory, Device, and Applications
http://dx.doi.org/10.5772/intechopen.110931
Edited by Yao-Feng Chang

Contributors
Amirali Amirsoleimani, Amrita Chakraborty, Deepthi Anirudhan Jayadevi, Guodong Li, Hassen Aziza, Jack Cai, Jiaming Wu, Marcelo Rozenberg, Marius K. Orlowski, Mohammad S. Al-Mamun, Muhammad Ahsan Kaleem, Nikolay Frick, Olaseinde Kehinde Femi, Roman Genov, Si Wu, Sourav Roy, Tianxian Zhang, Wanying Dai, Xiangliang Xu, Yao-Feng Chang, Yongqi Zhao

First published in London, United Kingdom, 2024 by IntechOpen
IntechOpen is the global imprint of INTECHOPEN LIMITED, registered in England and Wales, registration number: 11086078, 167-169 Great Portland Street, London, W1W 5PF, United Kingdom

British Library Cataloguing-in-Publication Data
A catalogue record for this book is available from the British Library

Additional hard and PDF copies can be obtained from orders@intechopen.com

Memristors - The Fourth Fundamental Circuit Element - Theory, Device, and Applications
Edited by Yao-Feng Chang
p. cm.
Print ISBN 978-0-85466-167-1
Online ISBN 978-0-85466-166-4
eBook (PDF) ISBN 978-0-85466-168-8

For EU product safety concerns:
IN TECH d.o.o., Prolaz Marije Krucifikse Kozulić 3, 51000 Rijeka, Croatia,
info@intechopen.com or visit our website at intechopen.com.

Meet the editor

Dr. Yao-Feng Chang is a part of the Technology and Development, Quality and Reliability (TD Q&R) staff at Intel. He received a Ph.D. from the Department of Electrical and Computer Engineering, University of Texas at Austin, USA, in 2015. His primary research focuses on emerging electronics and memory devices for advanced technology nodes for storage, computation, and energy-efficient integrated systems. He has published more than 100 journal publications and conference proceedings as well as five patents.

Contents

Preface

This book presents interdisciplinary research on memristor devices in fields ranging from physics to electrical analysis. First, it presents memristor studies covering basic theory, modeling, simulation, and fundamental two-terminal circuit elements. Then, the book moves on to discuss research at the device level, focusing on resistive random-access memory (ReRAM), phase-change memory (PCM), magnetoresistive random-access memory (MRAM), and ferroelectric random-access memory (FeRAM). The book also discusses memristor mechanisms and their corresponding electrical performance. Finally, the book highlights memristor system-level applications and hardware electrical applications, such as neuromorphic computing, conventional neural network/spiking neural network (CNN/SNN) memristive hardware computing, memory in processor, and processing in memory (PIM), as compared to conventional hardware systems, specifically power-performance-area-cost (PPAC) analysis.

Yao-Feng Chang
Intel Corporation,
United States of America

Section 1

Theory and Modeling

Chapter 1

Unveiling the Fourth Fundamental Circuit Element and Its Real-World Applications

Olaseinde Kehinde Femi

Abstract

Memristors are a type of electronic circuit element that was first proposed in the early 1970s. Unlike traditional circuit elements such as resistors, capacitors, and inductors, memristors has last memory and can therefore be used to store information. They were initially considered a theoretical concept, but recent advances in nanotechnology have made it possible to create physical memristor devices and apply it in various aspect of life. Memristors are considered the fourth fundamental circuit element, alongside resistors, capacitors, and inductors. They have unique properties, such as the ability to store information in their resistance state, which makes them promising candidates for future computing systems. Memristors have been integrated into crossbar arrays, which allow for massively parallel computing with low power consumption. Memristor fabrication methods vary based on the materials used and the intended application. Thin-film deposition, nanoimprint lithography, and self-assembly processes are common techniques. The performance properties of the memristor, such as switching speed, endurance, and scalability, are influenced by the material selection, such as polymers or transition metal oxides. Memristors have a wide range of potential applications, including in the development of artificial intelligence and neural networks. They can also be used in memory devices, logic circuits, and sensor applications. Research is ongoing to further explore the unique properties of this lovely device and to develop various applications for it. The unique properties of memristors have also sparked interest in unconventional computing paradigms. Memristor-based systems have shown potential for implementing neural networks, cellular automata, and even analog computers, providing alternative approaches to solving complex computational problems. Memristor-based logic and arithmetic units offer advantages in terms of power efficiency and density compared to traditional transistor-based designs. This project shows the progress made in memristor technology, including the development of various memristive materials and device architectures. It explores the challenges associated with memristor fabrication, reliability, and scalability. Moreover, the paper highlights recent advancements in memristor-based applications, such as in-memory computing, deep learning accelerators, and brain-inspired computing systems. This piece provides an overview of the theory behind memristors, including their mathematical models and properties. It also discusses the different types of memristor devices that have been developed and their potential applications. Finally, it highlights some of the challenges and future directions in the field of memristor research.

Keywords: memristor, electronics, computing, fourth, neural network, application, fabrication, nanotechnology, resistance, metal oxides, polymers

1. Introduction

Memristors are electronic devices and the fourth fundamental circuit element that were first proposed by Leon Chua in 1971 and later realized experimentally in 2008. They are characterized by their ability to remember the amount of charge that has passed through them, similar to the way a resistor remembers the amount of current that has passed through it. They have various potentials in many fields such as signal processing, artificial intelligence and computer Memory [1, 2].

The conventional Chua's diode has been suggested and implemented by several two-terminal passive electronic devices [3, 4]. Memristors are part of a group of devices known as "non-volatile" memory, meaning that they can store data which can be retrieved easily and even when power is removed from the device. The relationship between electronic charge q and the magnetic flux (∅), generally refers to as memristor W (∅) which has been realized physically by Stanley William's group from HP labs as the cubic-nonlinear function can be used for theoretical, numerical and experimental studies. The smooth cubic flux-controlled memristor can be characterized by a smooth continuous cubic nonlinearity given by:

$$\begin{cases} q(\phi) = \beta\phi + \xi\phi^3 \\ W(\phi) = \dfrac{dq(\phi)}{d\phi} = \beta + 3\xi\phi^2 \end{cases} \qquad (1)$$

and the current passing through the memristor is

$$i = W(\phi)v \qquad (2)$$

where β and £ are parameters embedded in the memristive-based function, and (∅) is the internal state variable of the smooth flux-controlled memristor. Also note that v is the input voltage, while i = f (W; v2) is the output voltage of the memristor. W(∅) is the conductivity of the memristor. The state variables correspond to voltages across each capacitor used in the designed circuit i.e. x = v1, y = v2, z = v3, w = (∅), v = v2 and the smooth cubic flux-controlled memristive function f(w) = W (∅). It is worth noting that w is the internal state of the memristive device [2] (**Figure 1**).

I will be considering the theory, devices, and applications of memristors below.

2. Theory

Leon Chua created the theory of memristors in 1971 after investigating other basic circuit components outside the resistor, capacitor, and inductor. Chua proposed the memristor, which has a unique characteristic known as memristance, as the fourth essential circuit component. A combination of the word's "memory" and "resistor," "memristor" emphasizes the device's capacity to store a recollection of the charge that has passed through it.

Memristors are identified by their uniqueness, known as the ability to store data, which is called "Memory". When a voltage is supplied across a memristor, the

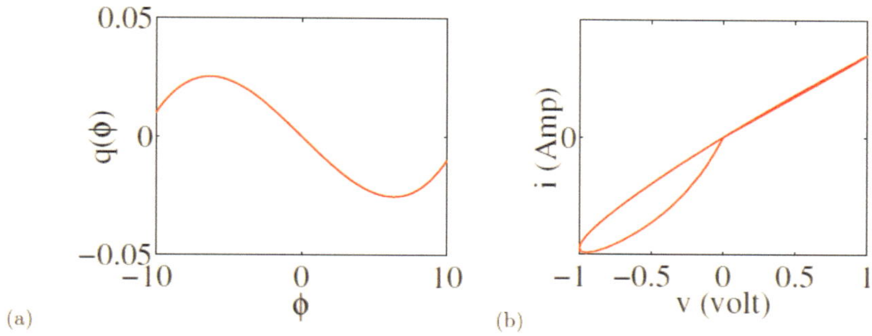

Figure 1.
Memristor model: (a) the relationship of charge-magnetic flux; (b) the i-v characteristic.

resistance of the device gets adjusted, and this change is sustained even when the voltage is removed. This property is referred to as "memristance," which is a combination of the word "memory" and "resistance." Memristors can be modeled using a mathematical equation such as a nonlinear dynamics system [5], which is similar to the equation that describes the behavior of capacitors, inductors, and resistors. This equation can be used to design and analyze memristor circuits.

At its core, the theory of memristors revolves around the relationship between charge (q), current (i), and flux (Φ). In traditional circuit elements, such as resistors, capacitors, and inductors, these variables are linearly related. However, the memristor introduces a non-linear relationship between them, allowing for the storage of information in the form of resistance.

The fundamental equation that defines the behavior of a memristor is:

$$v = M(q) * I \tag{3}$$

where v represents the voltage across the memristor, q is the charge that has passed through it, i is the current flowing through it, and M(q) is the memristance function.

The memristor's resistance is determined by the quantity of charge that has flowed through it and is described by the memristance function, M(q). The memristor's resistance may alter over time if the function has a temporal component. This characteristic is referred to as "dynamic memristance."

The term "flux linkage," shown by, describes the connection between charge and memristance. The integral of the voltage across the memristor with respect to time is represented by flux linkage:

$$\Phi = \int v\, dt \tag{4}$$

The memristance is determined by the derivative of the flux linkage with regard to charge:

$$M(q) = d\Phi/ \tag{5}$$

This derivative determines how the memristor's resistance changes as charge passes through it. If the memristance increases as the charge increases, the memristor is said to have "positive memristance." Conversely, if the memristance decreases with increasing charge, it is referred to as "negative memristance."

Memristors have the remarkable ability to maintain resistance levels even in the absence of power, making them non-volatile memory components. This feature results from the fact that total charge flow through the memristor, rather than momentary voltage or current, determines its state.

Pinched hysteresis loops shown in the current-voltage (I-V) characteristics provide as another illustration of the behavior of memristors. When a voltage is applied across a memristor, the current on the I-V graph follows a path that resembles a compressed figure-eight. The charge-dependent resistance leads to these squeezed hysteresis loops, which are a distinguishing characteristic of memristive behavior.

One of the key properties of memristors is their ability to retain their resistance state even when power is turned off, making them non-volatile memory elements. This arises from the fact that the memristor's state is determined by the total charge that has flowed through it rather than the instantaneous voltage or current. The memristor "remembers" its resistance state because it retains the accumulated charge.

The behavior of memristors can be further characterized by their "pinched hysteresis loops." When a voltage is applied across a memristor, the current flowing through it traces a loop-like path on the current-voltage (I-V) graph. This loop shape resembles a pinched figure-eight, hence the term "pinched hysteresis loop." The pinched hysteresis loop is a distinctive signature of memristive behavior and is a result of the charge-dependent resistance [6].

The theory of memristors has led to extensive research in both understanding their fundamental properties and developing practical applications. Memristors have the potential to revolutionize memory technologies, neuromorphic computing, analog circuitry, and more. As researchers delve deeper into the theory and fabrication of memristors, their applications are expected to expand, further unlocking their transformative capabilities in various fields of electronics and computing [7].

3. Devices

Memristors have been fabricated and proposed using a variety of materials such as resistive random-access memory (RRAM), including oxides, polymers, and nanowires. One of the most promising applications of memristors is in the development of "neuromorphic" computing systems, which a replica of the structure and function of the human brain Memristors can also be used in non-volatile memory devices, such as flash memory and solid-state drives, where they offer advantages over traditional memory technologies in terms of speed, space, power consumption, and durability [3–6, 8].

A solid-state electrical component with memristive behavior is known as a memristor device. It is based on the idea of a memristor, a passive, two-terminal electronic component whose resistance can vary depending on the amount of charge that has flowed through it. Due to their distinctive qualities, including non-volatility, great scalability, and prospective uses in memory, computation, and neuromorphic systems, memristor devices have drawn a lot of attention. A thin film of a memristive material is often deposited between two electrodes using thin-film deposition processes, such as sputtering or atomic layer deposition. The application requirements and desired performance dictate the choice of memristive material. Titanium dioxide (TiO2), hafnium oxide (HfO2), or tantalum oxide (Ta2O5) are examples of transition metal oxides or delete data in a non-volatile way.

Memristor technology has a number of benefits over conventional memory technologies. Their non-volatile nature, which enables them to maintain their resistance states even when the power is switched off, is one important advantage. Due to this property, they are suitable for in-memory computing and storage-class memory systems, which require quick and high-density non-volatile memory. Scalability, endurance, and low-power operation are all advantages of memristor-based memory devices like conductive-bridging random-access memory (CBRAM) and resistive random-access memory (RRAM).

Memristor technology also has a lot of potential for neuromorphic computing. Building artificial neural networks is attracted to them due to their capacity to imitate certain synaptic behaviors, including as long-term potentiation and spike-timing-dependent plasticity. Neuromorphic systems based on memristors have the potential to be parallel and energy-efficient [7, 9].

Pattern recognition, machine learning, and cognitive computing can be made possible by computing, through the use of memristor.

Memristor device research is advancing quickly, with continual efforts to enhance its functionality, dependability, and compatibility with current semiconductor technology. Future developments in memristor technology may result in innovations in memory, computation, and brain-inspired systems, influencing the design of electronic devices and systems in the future [10, 11].

The working principle of a memristor device is based on resistive switching, which refers to the ability to change the device's resistance by applying an electric field. Memristors can exist in two or more resistance states, typically referred to as high resistance (HRS) and low resistance (LRS). The transition between these states is non-volatile, meaning that it persists even when the power is turned off. The resistance of a memristor can be controlled by applying voltage pulses of different magnitudes and durations, allowing for precise manipulation of its electrical properties [9]. Memristor technology has also shown promise in analog and mixed-signal circuits. They can be used to build chaotic circuits, nonlinear dynamical systems, and analog signal processing operations. Memristors' capacity to store and retrieve analog states creates opportunities for the creation of novel circuit topologies, such as adaptable filters and reconfigurable analog circuits [12].

4. Applications

Memristors have various potential applications in a variety of fields, including computing sensing, neuromorphic computing, signal processing, artificial intelligence and energy storage. In Signal Processing, memristors have been used to realize non-linear-signal signal processing functions such as neural thresholding and spike base computation. In computing, memristors can be used to develop more efficient and powerful computer architectures, such as neural networks and deep learning algorithms. In sensing, memristors can be used to detect changes in electrical signals, which has applications in areas such as biosensing and environmental monitoring. In energy storage, memristors can be used to store energy, which has potential applications in areas such as renewable energy systems and electric vehicles [3, 8, 13–16].

With its distinctive electrical characteristics, memristor devices have created intriguing new opportunities for a variety of applications in electronics and computing. This study gives a thorough description of the various uses of memristors, emphasizing their potential to transform analog circuitry, neuromorphic computing, memory technology, and other fields. The study explores the particular applications, going into detail about their importance and possible impact. In addition, pertinent citations are included to back up the topic and offer further directions for investigation.

Few of the applications of memristors are listed below:

a. Non-volatile memory technologies: Memristors hold great promise as alternatives to conventional flash memory. This section discusses resistive random-access memory (RRAM) and crossbar arrays as examples of memristor-based memory technologies. The advantages of memristor-based memories, such as high-density storage, low power consumption, and compatibility with existing fabrication processes, are discussed in detail. Relevant citations showcase the research advancements and potential future developments in this field.

b. Neuromorphic computing: Memristors exhibit properties similar to synapses in biological neurons, making them ideal for implementing neuromorphic computing systems. This section explores the utilization of memristors in artificial neural networks and discusses their potential for energy-efficient, parallel computing. The concept of synaptic plasticity and spike-timing-dependent plasticity is explained, emphasizing the role of memristors in emulating these behaviors. Citations highlight notable research in neuromorphic computing using memristors and provide insights into future directions.

c. Analog and mixed-signal circuits: Memristors offer exciting opportunities in the field of analog and mixed-signal circuits. This section explores their potential applications in non-linear dynamical systems, chaotic circuits, and analog signal processing. The ability of memristors to store and recall analog states opens up new avenues for developing novel circuit architectures. Relevant citations illustrate the potential of memristors in these applications and showcase the progress made in implementing memristor-based analog circuits.

d. Beyond electronics and computing: This section discusses potential applications of memristors beyond traditional electronics and computing. It explores interdisciplinary areas where memristors can make a significant impact, such as bioelectronics, neuromorphic robotics, and brain-computer interfaces. Citations provide insights into the ongoing research in these interdisciplinary fields and highlight the potential future impact of memristors.

5. Conclusion

In conclusion, memristor technology is a tremendous advancement in computers and electronics with a wide range of revolutionary uses. Unlocking the full potential of memristor technology and advancing analog circuitry, computing systems, memory technologies, and multidisciplinary fields depend on ongoing research and development in this field. Memristors are expected to play a significant role in determining the direction of electronics and computing as the industry develops by providing novel solutions and pushing the envelope of what is practical.

Memristors are a promising area of research for the development of future computing technology. They offer unique properties that make them well-suited for a variety of applications, including neuromorphic computing, non-volatile memory, sensing, and energy storage. As researchers continue to explore the potential of memristors, we will likely see more applications of these devices soon (**Table 1**).

	Previous	Current
Theory	The earlier memristor theory, proposed by Leon Chua in 1971, introduced the concept of a fourth building block of a passive circuit alongside resistors, capacitors and inductors. Chua theorized that a memristor, short for "memory resistance," would possess a unique property called memristance. According to this theory, memristance represents memory-like behavior in resistive devices, where their resistance can be changed and "stored" depending on the magnitude and direction of the applied voltage. In other words, the memristor would have a different resistance depending on the amount of charge that previously flowed through it.	Memristors continue to be an active area of research and development, with ongoing efforts to understand their behavior, optimize their properties, and explore their applications in various fields
Devices	First Practical Memristor: After several decades, in 2008, researchers at Hewlett-Packard (HP) Labs led by Stan Williams successfully created the first practical memristor device. They used a thin film of titanium dioxide as the active material, sandwiched between two metal electrodes. The resistance of the device could be adjusted by applying voltage across it.	There are Improved Materials and Variations, which researchers continued to explore different materials and structures to improve the performance of memristors. Various oxides, such as hafnium oxide, tantalum oxide, and niobium oxide, were tested as potential active materials. These advancements led to improved device characteristics, such as faster switching speed, lower power consumption, and higher endurance.
Application	Here are some previous applications of memristor as stated below: Non-Volatile Memory: One of the primary applications of memristors is in non-volatile memory technology. Memristors can retain their resistance state even when power is turned off, making them suitable for use in high-density and energy-efficient memory devices. Neuromorphic Computing: Memristors have gained significant attention in the field of neuromorphic computing, which aims to build brain-inspired computing systems Reconfigurable Circuits: Memristors can be utilized in reconfigurable circuits, enabling the creation of programmable analog and digital circuits. By adjusting the resistance of memristors, the behavior and functionality of the circuits can be modified dynamically, allowing for adaptive and flexible circuit designs.	Memristors are still in the early stages of commercialization, and their widespread adoption is growing rapidly in all areas such as: Artificial Intelligence (AI) Hardware Acceleration: Memristors hold potential for accelerating AI computations by providing efficient and scalable hardware solutions. Their ability to perform analog computations and emulate synaptic behavior can enhance the efficiency of neural network training and inference tasks, leading to improved AI performance. Bio-Inspired Systems: Memristors are being investigated for their potential in bio-inspired systems, such as brain-machine interfaces and neuro prosthetics. The ability to emulate synaptic behavior and process analog signals makes memristors suitable for building interfaces between electronic systems and biological organisms.

	Previous	Current
	Sensor and IoT Applications: Memristors can be integrated into sensing devices to enable novel functionalities.	It's important to note that the field of memristor research is still evolving, and new applications and advancements continue to emerge. Researchers and industry experts are actively exploring the potential of memristors and working towards their integration into practical applications across various spheres.

Table 1.
Simple bench mark table showing previous and current work on memristor: Theory, device and application.

Author details

Olaseinde Kehinde Femi
Federal University of Technology Akure, Akure, Nigeria

*Address all correspondence to: olaseindekehindefemi@gmail.com

References

[1] Chua LO. Memristor—The missing circuit element. IEEE Transactions on Circuit Theory. 1971;**18**(5):507-519

[2] Strukov DB, Snider GS, Stewart DR, Williams RS. The missing memristor found nature. 2008;**453**(7191):80-83

[3] Bao BC, Jiang P, Wu HG, Hu FW. Complex transient dynamics in periodically forced memristive Chua's circuit. Nonlinear Dynamics. 2015; **79**(43):2333

[4] Wang G, Yuan F, Chen G, Zhang Y. Coexisting multiple attractors and riddled basins of a memristive system. Chaos. 2018;**21**(1):013125. DOI: 10.1063/1.5004001

[5] Zhang Q, Zhu J, Chen L, Wu H. A survey on the modelling of memristor devices. IEEE Access. 2019;7:8511-8522

[6] Battistoni S. Organic memristive devices for neuromorphic applications. Bio Nano Science. 2021;**11**:227-231. DOI: 10.1007/s12668-020-00808-z

[7] L. O. Memristor and memristive systems. In: Tetzlaff R, editor. Proceedings of the IEEE. 2011;**100**(6): 1859-1870. DOI: 10.1109/JPROC.2011.2151650

[8] Chen M, Li MY, Yu Q, Bao BC, Xu Q, Wang J. Dynamics of self-excited attractors and hidden attractors in generalized memristor-based Chua's circuit. Nonlinear Dynamics. 2015;**81**:215

[9] Yang JJ, Strukov DB, Stewart DR. Memristive devices for computing. Nature Nanotechnology. 2013;**8**(1): 13-24. DOI: 10.1038/nnano.2012.240

[10] Ielmini D, Wong HSP. In-memory computing with resistive switching devices. Nature Electronics. 2018;**1**(6): 333-343. DOI: 10.1038/s41928-018-0092-8

[11] Prezioso M, Merrikh-Bayat F, Hoskins BD, Adam GC, Likharev KK, Strukov DB. Training and operation of an integrated neuromorphic network based on metal-oxide memristors. Nature. 2015;**521**:7550

[12] Wong H-SP, Lee H-Y, Yu S, Chen Y-S, Wu Y, Chen P-S. Metal–oxide RRAM. Proceedings of the IEEE. 2012;**100**(6): 1951-1970. DOI: 10.1109/JPROC.2012.2187245

[13] Fitch AL, Yu DS, Herberth HCI, Sreeram V. Hyperchaos in a memristor-based on modified canonical Chua's circuit. International Journal of Bifurcation Chaos. 2012;**22**(6):1250133

[14] Yang JJ, Borghetti J, Murphy D, Stewart DR, Williams RS. A roadmap for nanoelectronics enabled by memristor devices. Nature Nanotechnology. 2013; **8**(10):731-734

[15] Bao BC, Hu FW, Liu Z, Xu JP. Mapping equivalent approach to analysis and realization of a memristor-based dynamical circuit. China Physics B. 2014;**23**(7):303-310

[16] Indiveri G et al. Neuromorphic silicon neuron circuits. Frontiers in Neuroscience. 2011;**5**:73. DOI: 10.3389/fnins.2011.00073

Chapter 2

A Trivial Implementation of an Analog Spiking Neuron Using a Memristor, for Less than $1

Jiaming Wu and Marcelo Rozenberg

Abstract

Memristors are finding widespread applications in neuromorphic circuits due to their unique resistance memory effect. Nonvolatile memristors are used for implementing artificial synapses and volatile ones for spiking neurons. An important example of the latter is the memristive neurons based on Mott insulators. However, fabricating and understanding volatile memristors based on Mott materials remains a difficult challenge, which hinders their adoption. In recent years, we have been developing an alternative approach based on a novel volatile device that is trivially made with a thyristor and a resistor. These two ordinary out-of-the-shelf conventional electronic components make our memristive device trivial to implement, widely available, reliable and extremely affordable. The key nontrivial insight was to recognize that it exhibits a memristive current-voltage characteristic qualitatively identical to that of Mott insulators. Here, we introduce in detail our device and show how it can be used to implement spiking neurons. We discuss the example of a bursting-neuron circuit model, which exhibits spiking behaviors in remarkable agreement to some observed in biological bursting neurons of mammals. The simplicity and low cost of our neuromorphic hardware makes it an ideal platform for implementing electroceutical medical devices for neuropathologies like epilepsy and Parkinson's disease.

Keywords: spiking neuron, neuronal networks, artificial intelligence, bursting, thyristor

1. Introduction

Since the theory of memristors was proposed in the 1970s [1, 2], these devices have been regarded as the fourth fundamental circuit element and have attracted much attention. However, that theory has also provoked some debate [3]. Besides that debate, the physical phenomenon of a resistance, whose value depends on the past history of the current that has circulated thought it, is well established. The potential technological interest has been the initial driver for the massively increasing number of publications that are produced [4]. Memristors have been already applied in various areas, such as resistive random access memories (ReRAM), in-memory computing

(IMC), and neuromorphic hardware [5–8]. The application in neuromorphic technology has gained particularly widespread attention in the AI. This is motivated by the possibility of using memristive arrays to process massive data at extremely low power consumption [9–11].

Neuromorphic computing is a computational approach inspired by the structure and function of the neural network of human brain [12, 13]. The basic components of neural networks are neurons that are interconnected by synapses [14, 15]. The former can be viewed as computational units that use the information that is stored in the latter. Hence, there is intimate physical proximity between CPU and memory, which is in drastic contrast with the traditional von Neumann computer architecture. In neuromorphic hardware, nonvolatile memristors are usually used to implement the functions of synapses [16–19], while volatile memristors are used to implement the artificial neurons [20–23]. One of the most prominent types of materials that are adopted for fabricating memristors are transition metal oxides [24]. Some of the most popular nonvolatile resistive switching compounds are TiO2, HfO2, and Ta2O5, which are finding their way into large crossbar arrays [25]. On the other hand, volatile compounds for neuristors are NbO2 [26] and specially vanadates [24], such as VO2 [27], V2O3 [28], and V3O5 [29]. However, a major challenge that the implementation of this type of neuromorphic devices still faces is their lack of reliability and systematic reproducibility of their characteristics [20, 24, 30]. This issue is significantly more severe for the neuristors, that is, the electronic spiking neurons, which are based on Mott quantum materials. The key property of the so-called Mott insulators that allows them to implement spiking neurons is that they exhibit a spectacular insulator-to-metal transition as function of temperature. In essence, upon application of short and intense voltage pulses, the Mott insulator accumulates the effect of the electric stress until, suddenly, its resistance collapses. As it does, a surge of current flows though the device. That surge is assimilated to the emission of an action potential [31]. We should also mention that another aspect that complicates the situation further is that the theoretical description of Mott materials is also challenging [32]. This is even more so when the materials are pushed away from equilibrium, under the action of strong electric pulsing. Therefore, to provide qualitative understanding of the physics, phenomenological descriptions are a useful [23]. Thus, despite much current effort, the implementation of reliable Mott insulator devices for realizing artificial spiking neurons remains a significant problem of material science.

In this context, in this book chapter, we describe our recent efforts to propose a new volatile memristive device that permits a practical, simple, affordable, and reliable implementation of an electronic spiking neuron [33–35]. Our device is composed of inexpensive out-of-the-shelf components, namely, a thyristor and a resistor, which will be referred to as the "M-device" in the following (M for Marcelo). The key insight that enables this breakthrough was to realize that the thyristor has an I-V characteristic that closely resembles that of Mott insulator materials, such as VO2, NbO2, and GaTa4Se8 [21, 26, 36, 37]. However, the thyristor is a three-terminal device, while the Mott devices only one. Hence, a second key insight was to realize that analogy with the Mott I-V characteristics can be made complete by simply connecting the gate and anode terminals of the thyristor with a resistance. In **Figure 1** below, we show the qualitative agreement between the I-V characteristics of NbO2 and our M-device.

To fully demonstrate this full qualitative agreement between the Mott and the M-device behaviors, we show in the **Figure 2** below that both systems when connected in parallel with a capacitor produce the same type of periodic oscillations which resemble the periodic neuronal tonic spiking.

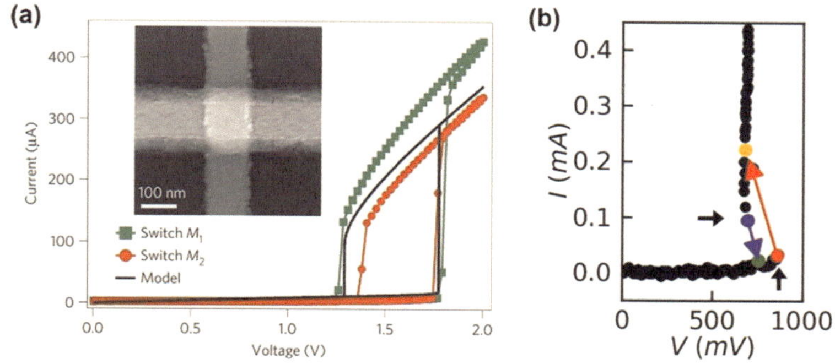

Figure 1.
(a) The current-voltage curves of two NbO2 crosspoint devices [reproduced from [26]]. (b) the I-V curve of the M-device displays the hysteresis effect as well, being qualitatively same as that of the NbO2 (the detail of the M-device and the measurement setting will be explained later in section 2, where ***Figure 4(a)*** *plots the same data as in this panel).*

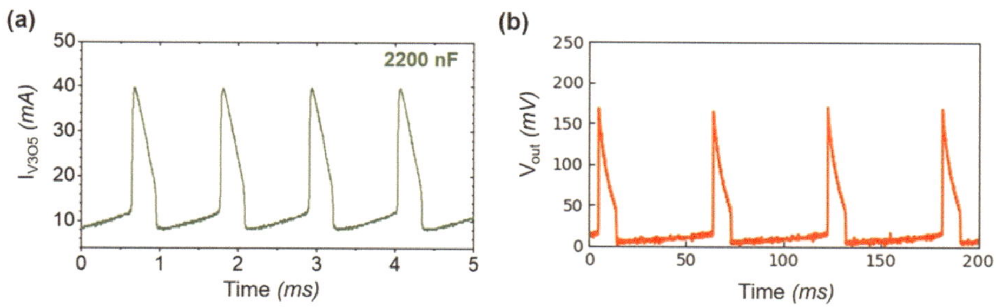

Figure 2.
(a) Time dependence of the current through the V3O5 Mott device [reproduced from [29]]. (b) the periodic oscillations measured in the M-device circuit (the y-axis is the output voltage, which is proportional to the current going through the M-device). The detailed circuit and measurement setting will be shown later in section 3, where the No. 3 example trace of ***Figure 9*** *plots the same data as in this panel.*

In the present work, we shall describe in detail how our M-device volatile memristor is a simple solution that allows us to build reliable artificial neuron circuits in a very cost-effective and straightforward manner. We shall first describe a basic memristive spiking neuron (MSN) and then extend it to implement a memristive spiking bursting neuron (MSBN) with more complex dynamical behaviors. We shall explore the full parameter space on the MSBN, obtaining a phase diagram of neuronal behaviors. Surprisingly, the phase diagram exploration allowed us to discover various qualitatively different spiking modes that bear striking similarities with those observed in real bursting neurons, rendering unexpected biological relevance to our neuron circuit model. This suggests exciting direction for future developments, such as the implementation of electroceuticals for the medical treatment of neurodegenerative diseases, such as Parkinson's [38] and epilepsy.

Since the thyristor is a conventional electronic device, one may also extend the memristive circuit to the realm of very-large-scale-integrated CMOS circuits (VLSI) [39]. This enables the fabrication of massive numbers of spiking neurons and adopt it as a platform of large artificial neural networks for Artificial Intelligence applications.

2. The memristive device based on the thyristor

The thyristor, also known as the Silicon-Controlled Rectifier (SCR), is a conventional semiconductor current control device invented in the 1950s [40]. It is composed of p-type and n-type materials arranged in a p-n-p-n four-layer structure. **Figure 3(a)** and (b) show the schematic and circuit symbol of the thyristor, which has three terminals: Anode, Cathode, and Gate (A, K, G).

The working principle of a thyristor can be most simply described as a diode with a threshold. A diode is a p-n junction that exhibits low resistance in the forward direction (say, p to n for positive bias) but experiences a significantly high resistance (low conductance) when subjected to a reverse voltage bias. Therefore, a positively biased thyristor can be regarded as two p-n junctions in direct (J1 and J3 in **Figure 3(a)**), and one in inverse (J2). Therefore, when a forward voltage is applied between Anode and Cathode, and no external voltage applied to the Gate, J1, and J3 would be able to

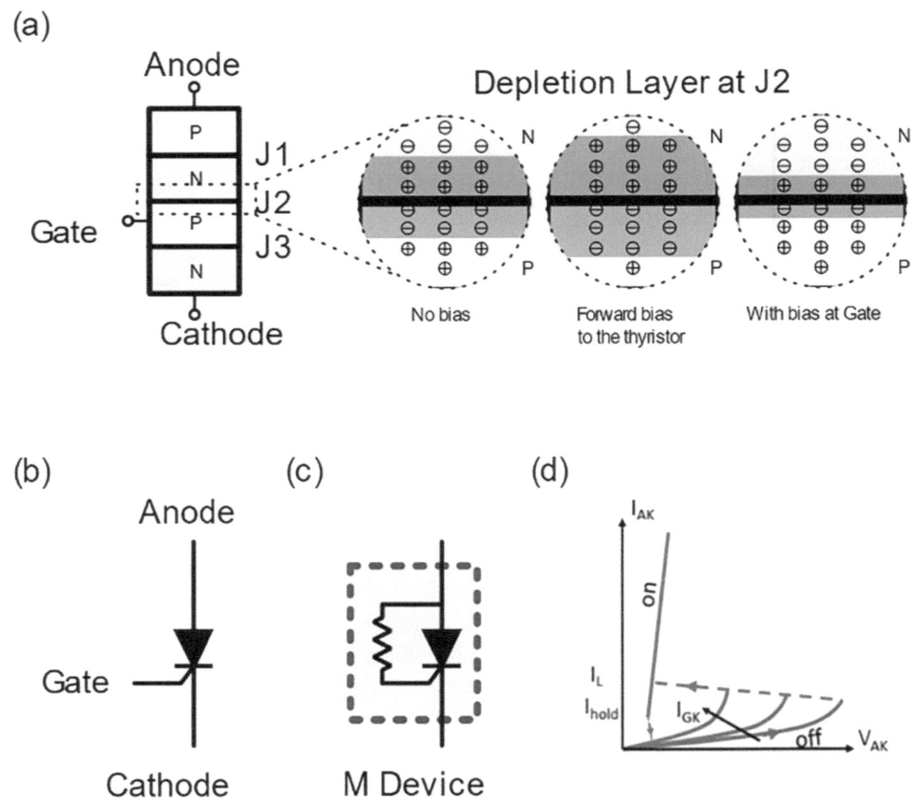

Figure 3.
(a) Schematic of the thyristor with p-n-p-n four-layer structure, and the changes of the width of the depletion layer at the J2 junction under different external conditions. (b) Circuit symbol of the thyristor. (c) Schematic of the M-device, consists of one thyristor and one resistor connecting the anode and gate of the thyristor. (d) Characteristic I-V of a thyristor in forward bias. Initially, the thyristor is in the high-resistance sate as there is a depletion layer in the inverse junction J2. The gate current (or applied voltage) controls the threshold switch-on value. Note that the I-V displays a negative differential resistance, which is key for the excitability of the memristive neuron device. In addition, there is also hysteresis, which is crucial to produce the spiking *behavior in contrast to simple periodic oscillations (see section 3) [(d) is adapted from [34]].*

conduct. However, J2 being reverse-biased creates a large depletion region that is difficult for carriers to overcome (**Figure 3(a)**). Thus, the thyristor is at a very low conductance state. By applying bias to the Gate, the electrons/holes forming the depletion region can be removed so that the depletion layer becomes thinner (**Figure 3(a)**). Thus, there are in principle two ways to switch on the device, one by applying sufficient positive bias so that a current is forced through the depletion region (as in an inverse bias breakdown of a diode) or by injecting current into the gate. Indeed, when the Gate current (or voltage) reaches a sufficiently high level, the depletion layer becomes "flooded," and the device switches to the high conductive state. These two phenomena are intertwined, so we can think of a threshold AK voltage that is controlled by the GK current (see **Figure 3(d)**).

Based on this property of the thyristor, a key insight is that one may connect the Gate directly to the Anode through a resistor R_{gate} so as to form a two-terminal device that we shall call "M-device" (**Figure 3(c)**). In this way, without the need for an additional control of the Gate bias, when the (external) voltage is increased at the Anode, the voltage (and current) at the Gate will also increase. Thus, by controlling the value of the R_{gate}, we may tune the resistive switching I-V characteristics of the two-terminal device (**Figure 4(b)**). When applied voltage V_{AK} is small, the Gate voltage V_{GK} (and I_{GK}) is also small; then, the M-device is at high resistance as the depletion layer of the thyristor is in place. As the voltage V_{AK} increases, the current injected into the J2 through the Gate also increases so that the depletion layer starts to thin out. At a certain applied voltage, the M-device reaches a threshold where the depletion layer suddenly collapses and the AK resistance of the M-device commutes, switching from the high resistance (low conductance) to the low resistance (high conductance) (**Figure 4(a)** blue arrow).

Another key property worth noting of the M-device characteristics is the hysteresis, which originates in the "holding" current of the thyristor. Once the thyristor is at a high conductance state, a large current density is flowing through the formerly depleted region. Hence, even the V_{AK} is falling below the threshold, the high conductive state remains self-sustained as the current density prevents the depletion layer to reform. As the V_{AK} is further decreased, the current density through the M-device also decreases. Eventually, that current density may decrease below small value called

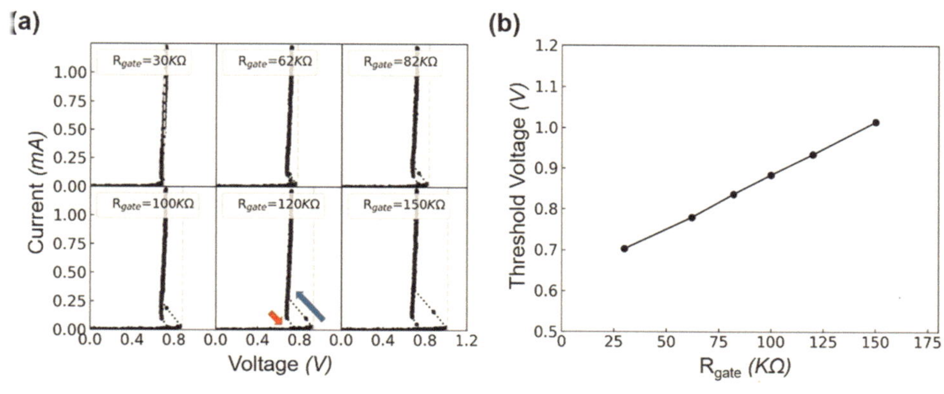

Figure 4.
(a) The measured I-V characteristics of the M-device with different R_{gate} values. (b) Linear relation between the R_{gate} and the threshold voltage of the M-device (the I-V curves are measured with a 1KΩ resistor in series with the M-device and driven by voltage wave generator. The thyristors used in the experiments are the SCR from STMicroelectronics with reference number P0118MA 2AL3).

Component	Brand	Manufacturer part number	Cost
Thyristor	STMicroelectronics	P0118MA 2AL3	~0.3 $
Resistor	Nova	CBR-14	<0.1 $

Table 1.
The cost of materials to make one M-device. The cost is estimated according to common suppliers on the market.

the holding current (I_{hold}) where the depletion layer forms again. This produces the switching back of the M-device to the low conductance state (**Figure 4(a)** red arrow).

These mechanisms enable the memristive characteristic of the M-device even with different R_{gate}, which is shown in the measured I-V curves in **Figure 4(a)** [33]. They perform the behavior of "pinched" hysteresis loops [41]. For instance, in the bottom middle panel of **Figure 4(a)**, the blue arrow indicates the transition from the low-conductance state to the high-conductance state at a threshold voltage, and the red arrow indicates the reverse transition at I_{hold}. This I-V curve carries qualitative similarities to the I-V loops of volatile Mott insulators [36].

In addition, from the engineering point of view, the threshold voltage (V_{th}) for the M-device to switch from high resistance to low resistance, a key parameter of the memristive spiking neuron model described in the next section, is easily adjustable. **Figure 4(a)** and **(b)** show that by increasing the value of the R_{gate}, the V_{th} also increases correspondingly. Moreover, there exists a linear relationship between the R_{gate} and the V_{th}, granting it with excellent engineering controllability.

As the thyristor and resistor are both stable, inexpensive, and out-of-the-shelf electronic components, they make the M-device a reliable and cost-effective volatile memristive device. The cost of building an M-device is much less than 1 $, as listed in **Table 1**.

3. The memristive spiking neuron model

In this section, we describe the implementation of an analog spiking neuron model exploiting the volatile memristive properties of the M-device. We shall first introduce a basic circuit configuration that implements a leaky-integrate-and-fire model neuron. We shall then extend the basic model to implement a bursting neuron with rich dynamic characteristics. An important point to make is that while a mathematical model is defined by a set of equations, the present neuron models are defined by their hardware implementation. In other words, *the circuit is the model*.

A typical biological neuron is schematically represented in **Figure 5(a)**. The neuron receives input current signals through the dendrites and generates electric action potential pulses (spikes) that travel down the axon and generate the output to other (downstream) neurons. A molecular-level mechanism of the spiking neuron can be roughly described by looking at the membrane of the neuron cell depicted in **Figure 5 (b)**. The cell body of the neuron is surrounded by a membrane, which is a lipid bilayer containing various types of embedded protein structures [42]. The lipid bilayer membrane is an excellent insulator, preventing the free flow of ions (such as sodium and potassium ions) between the two sides of the cell membrane. However, there are also proteins that form many ion channels across the cell membrane to actively transport or passively allow the passage of ions under certain conditions. Therefore, from a

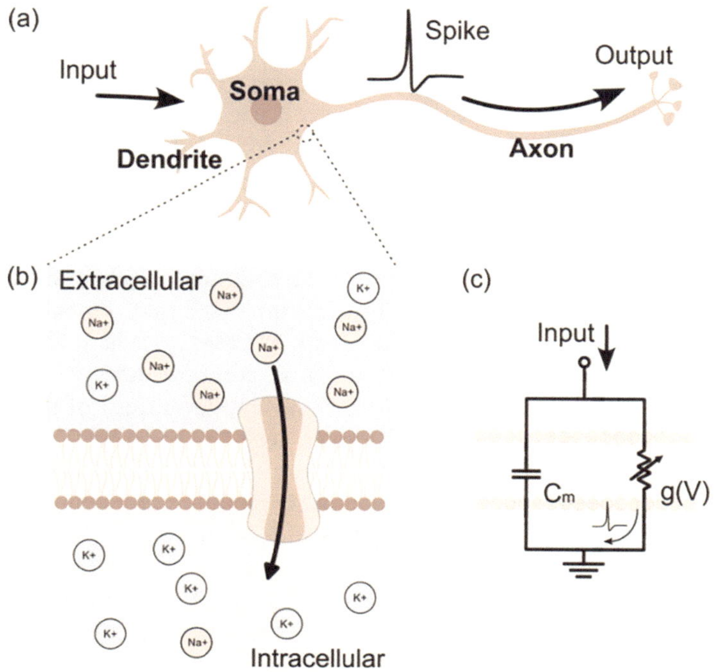

Figure 5.
Schematic of (a) a biological neuron, (b) its membrane structure with lipid bilayer and ion channels, and (c) an equivalent circuit of the integrate-and-fire model, where we only show one conductance channel for simplicity.

functional perspective, the lipid bilayer can be represented as a capacitor C_m, while the ion channels can be represented by several nonlinear conductances whose values depend on the membrane voltage $g(V)$. The latter are also called voltage-gated conductance channels, and in our neuron model, they are implemented by the M-device, as we shall discuss next.

The very simplified working principle of such a system can be described as an Integrate-and-Fire (IF) neuron model, as shown in the **Figure 5(c)**. The input currents to the neuron, the charges (ions in the biological case), are first "integrated" in the membrane capacitor C_m. The resulting voltage of the C_m (membrane potential V_m) increases until it reaches some certain threshold value, where the conductance channel $g(V)$ becomes conductive. Then, the C_m rapidly discharges through the conductance channel, that is, "fires" a strong current or voltage spike, which is called the action potential. This basic mechanism is also present in some more biologically realistic models such as in the Hodgkin-Huxley model or, in its simplified version, the Morris-Lecar model [15, 43–45], where multiple ion channels may be considered, with different voltage-gated channels represented by respective nonlinear conductances. Regardless of their simplicity or complexity, all those models share a similar mathematical structure, which is defined by a set of equations of the form:

$$C_m \frac{dV_m}{dt} = I_{ion}(t) + I_{in}(t) = \sum\nolimits_k I_k(t) + I_{in}(t) \tag{1}$$

$$I_k(t) = g_k(V_m, S_k)V_m\# \tag{2}$$

where I_ks are the channel currents and $g_k(V_m, S_k)$ is the respective voltage-gated conductances. The variable S_k indicates the "state" of the conductance.

In an innovative hardware implementation that is receiving significant attention, a type of quantum materials called Mott insulators is used to implement the nonlinearity required in a voltage-gated channels [20, 21]. The key insight in the present approach is, as demonstrated in Section 2, that the M-devices exhibit I-V characteristics that are analogous to those of Mott insulators, particularly in terms of their volatile history-dependent conductivity or memristance. Therefore, M-devices exhibit the key feature required to implement the conductance channels of **Eqs. (1)** and **(2)** above. Moreover, in contrast to the Mott materials that remain a challenge to reliably fabricate and to theoretically understand, our M-device is solely implemented in inexpensive conventional off-the-shelf electronic components.

Figure 6(a) shows the circuit implementation (i.e., definition) of a Memristive Spiking Neuron (MSN). As can be seen in **Figure 6(c)**, the basic MSN circuit realizes a (Leaky) Integrate-and-Fire neuron model. In this model, $\boldsymbol{C_m}$ plays the role of the lipid bilayer while the M-device acts as a (leaky) ion channel. For convenience, in order to transform the current spikes produced by the sudden discharge of the $\boldsymbol{C_m}$ into a voltage spike signal $\boldsymbol{V_{out}}$, we adopt a small $\boldsymbol{R_{load}}$ (47 Ω) in series with the M-device.

Under the excitation of a constant current $\boldsymbol{I_{in}}$, the MSN circuit emits a regular succession the electric spikes, which is called tonic spiking. One such a spike is shown in the top right panel of **Figure 6(b)**. During the spike generation, a conductance switch of the M-device is clearly observed, as shown in the main panel of **Figure 6(b)**. The colored dots with numbers indicate the conductance state of the M-device that are correlated to the respective colored dots along the spike trace in the top-right panel. When the M-device switches from low conductance to high conductance (dot 1 to dot 2), a significant increase in $\boldsymbol{V_{out}}$ is observed; that is, the spike is emitted. Then, the $\boldsymbol{C_m}$ rapidly discharges and the current going through the M-device decrease to $\boldsymbol{I_{hold}}$ (dot 2 to dot 3, and also see **Figure 4**). Finally, the M-device switches back to low conductance (dot 3 to dot 4). Note that the difference between slow charge time (integrate) and fast discharge (fire) corresponds to the dramatic change in the RC constant of the circuit due to the conductance change of the M-device.

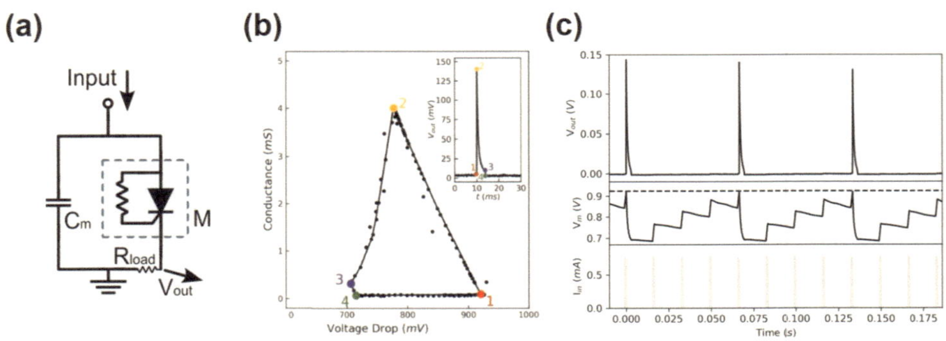

Figure 6.
(a) Schematic of the Memristive spiking neuron (MSN) circuit based on the M-device. (b) a typical spike generated by the MSN, and the corresponding nonlinear conductance changes of the M-device during the spike event [adapted from [33]]. (c) Demonstration of the MSN integrate-and-fire function under a train of input pulses. The mid panel shows the leaky integration of the current in the C_m. The dotted line indicates the threshold potential. When V_m reaches the threshold, a spike is emitted, as shown in the top panel. The bottom panel shows the train of input synaptic current pulses. The data are obtained with $C_m = 10\mu F, R_{gate} = 100K\Omega, R_{load} = 47\Omega$, and $I_{in} = 92.4\mu A$.

The Integrate-and-Fire behavior is also very intuitively observed when the input is a train of synaptic current pulses, as seen in **Figure 6(c)**. The pulse inputs are (leaky) integrated in the C_m one by one, until the membrane potential V_m reaches certain threshold; then, a spike is generated and V_m is reset. One important thing to note here is that during the time between two input pulses, there is no input current and the membrane potential shows a small decrease. This is the leaky integration, which is due to the fact that the M-device is not an ideal switch (i.e., the low conductance is not open circuit). The small charge leakage through the M-device is controlled by the value of R that is connected between the anode and the gate of the thyristor (see Section 2). Hence, the MSN circuit may be better characterized as a realization of the Leaky-Integrate-and-Fire (LIF) model, which will be further elaborated in the following discussion.

From this description, one can readily understand that the spike frequency (firing rate) should depend on the input current. For instance, upon the excitation with a constant current I_{in}, the membrane potential V_m of an ideal IF neuron linearly increases with time, as $V_m(t) = \frac{I_{in} \cdot t}{C_m}$, and suddenly discharges with a spike when $V_m(t) = V_{th}$. Therefore, an IF neuron model emits spikes at a frequency $f(I_{in}) = \frac{I_{in}}{C_m \cdot V_{th}}$, which is linear in the applied I_{in}. This behavior is qualitatively well reproduced by our MSN model, as shown in **Figure 7**.

However, it is worth noting that such a linear relationship between spiking frequency and I_{in} is not applicable across the entire I_{in} range. As shown in **Figure 7**, deviations from this simple linear relationship occur when I_{in} is too small or too large. In **Figure 7**, we denote by I_{in}^{min} and I_{in}^{max} the lower and upper thresholds of excitability; namely, they define the current interval where spike emission is triggered in the MSN model. Biological neurons also have qualitatively similar thresholds. The region beneath the lower one may be associated to the so called “hyperpolarized” state, while the region above the upper threshold may be associated to the “depolarized” one. Another consequence of these two boundaries is that there exists a maximal spike

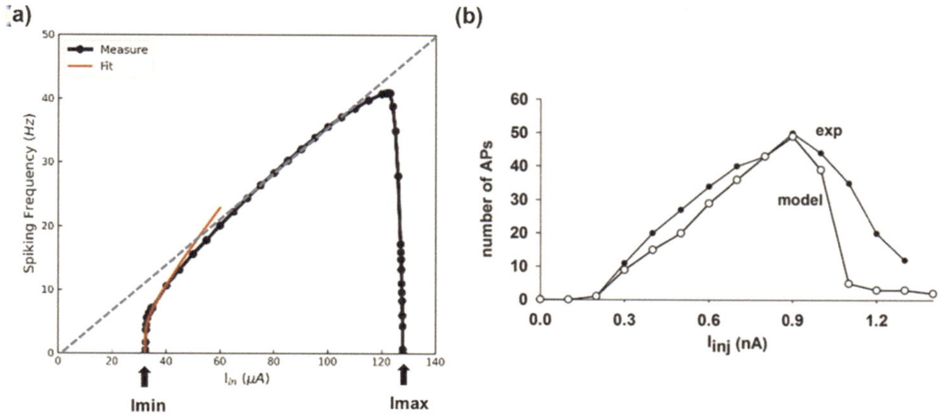

Figure 7.
(a) The spiking frequency's dependence on input current I_{in} of the MSN (measured with $C_m = 10\mu F$, $R_{gate} = 100K\Omega, R_{load} = 47\Omega$). In the middle region, the curve follows an approximate linear behavior (blue dashed line) similar to the ideal IF mode. Deviations from linearity occur near the excitation thresholds I_{in}^{min} and I_{in}^{max}. The red curve indicates the fitted frequency versus input relationship of an LIF model [45]. (b) the experimental data of the current dependent spike rate of a biological neuron (exp) that shows the same qualitative behavior of our MSN model. The sudden decrease of spike rate with increasing current is called the “depolarization block” [(a) is adapted from [33], and (b) is reproduced from [46]].

frequency rate that can be sustained by the neuron, also qualitatively similar to that present in biological neurons, as shown in **Figure 7(b)**. At the maximal frequency, we obtain a minimal inter-spike time interval that may be considered as a refractory period of the neuron.

The reason for the departure from the linear behavior at I_{in}^{min} is due to the leakage of the M-device. As explained before, the conductance of the M-device is not zero when it is at the low conductance state, which results in a leak current. Thus, the I_{in}^{min} is due to the minimal current I_{in} that is required to overcome the leakage so as to enable the charge integration in C_m and eventually emit a spike. In the LIF model, $f(I_{in})$ and I_{in} are no longer linearly related, but they follow a logarithmic relationship [15, 45]:

$$f(I_{in}) \propto -\frac{1}{\log\left(1-\frac{I_{in}^{min}}{I_{in}}\right)} \# \tag{3}$$

We observe in **Figure** 7 the good quality of the fit using the expression of the LIF model in **Eq. (3)**. Those data also demonstrate the type 1 excitability of the MSN model, which refers to systems that initiate spiking from zero frequency [15, 45].

As for the deviation from linear behavior at I_{max}, it is mainly related to the holding current. More specifically, it happens when the I_{in} is greater than I_{hold}. In this case, once the M-device switches to the high conductance state (dot 1 to dot 2 in **Figure 6(b)**), the C_m is continuously discharging, but the charge is replaced by the relatively large I_{in}. Thus, for $I_{in} > I_{hold}$, the M-device never gets the chance to switch back to its low conductance state (dot 3 to dot 4 in **Figure 6(b)**). The single spike is a phasic firing, which is followed by the neuron accommodating to a continuous flow of a large current through the M-device. This high-current state is reminiscent of the depolarization block observed in biological neurons (see **Figure** 7**(b)**) [46].

4. The memristive bursting neuron model

To search for more complex neuronal dynamics, we extended the MSN into a Memristive Spiking Bursting Neuron (MSBN) model by introducing a second time constant in the circuit model. This is inspired by the two-compartment models (**Figure 8(a)**) from theoretical neuroscience such as Pinsky-Rinzel model [47]. We introduced a second capacitor C_s to the MSN in parallel with the load resistor R_s as shown in **Figure 8(b)** The resulting MSBN model exhibits, in addition to the simple spiking behavior (referring to Tonic Spiking (TS) shown in **Figure 8**), three new spiking modes: Fast Spiking (FS) and two types of Intrinsic Bursting (IB Type1 and IB Type2) as displayed in **Figure 8**.

Thus, the MSBN has, in addition to the first time constant in the circuit defined by the $\tau_m = R_{gate}C_m$, a second time constant, $\tau_s = R_sC_s$. Then, similar to the Pinsky-Rinzel model [39], we associate each part of the circuit to two compartments, namely, a dendrite compartment and a soma compartment, respectively. The increase in dimensionality of this model is considered a necessary condition for generating intrinsic bursting [45]. More specifically, C_s and its voltage V_s introduce an additional modulation to the I_{ion} in **Eq. (1)**, leading to the following model equations:

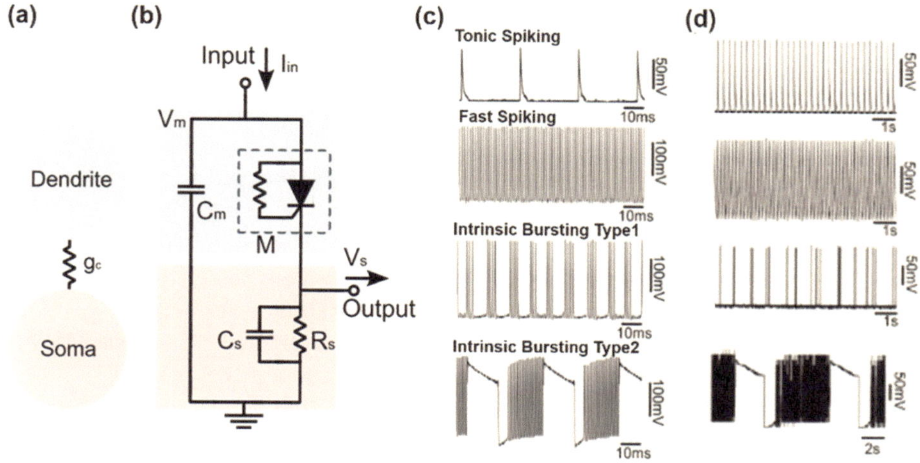

Figure 8.
(a) Schematic representation of a two-compartment neuron model. (b) Memristive spiking bursting neuron (MSBN) inspired by the two-compartment model. (c) Four neuronal behaviors generated by the MSBN, including tonic spiking, fast spiking, intrinsic bursting type 1 and type 2. (d) the same spiking and bursting traces generated in a biological realistic timescale by tuning the value of C_m. Measurement settings: in (c), TS is obtained at (C_m, I_{in}, R_s, C_s) = (10 μF, 72.2 μA, 47 Ω, 0.1 μF); FS at (10 μF, 46.7 μA, 2.4 kΩ, 0.1 μF), IB1 at (10 μF, 38.7 μA, 3 kΩ, 0.1 μF), IB2 at (10 μF, 75.1 μA, 2.4 kΩ, 0.1 μF). In (d), TS is obtained at (C_m, I_{in}, R_s, C_s) = (100 μF, 93.8 μA, 100 Ω, 0.1 μF), FS at (1000 μF, 54.1 μA, 3 kΩ, 33 μF), IB1 at (1000 μF, 29.4 μA, 6.8 kΩ, 0.47 μF), IB2 at (2000 μF, 76.2 μA, 2.3 kΩ, 0.1 μF) [adapted from [33]].

$$C_m \frac{dV_m}{dt} = I_{ion}(g, \tau_s) + I_{in} \quad (4)$$

$$I_{ion} = g(V_m, S)(V_m - V_s) \quad (5)$$

$$C_s \frac{dV_s}{dt} = I_{ion} - \frac{C_s}{\tau_s} V_s \quad (6)$$

Note that the last equation in this set shares the same form as the dynamic equation for the [Ca]-current in the theoretical model of Bursting neurons [45].

The MSBN has, nevertheless, some differences from those theoretical mathematical models. In those models, the second time constant τ_s is usually greater than τ_m so as to slowly modulate the tonic spike generation by driving the neuron in-and-out of its excitability range. In contrast, in the MSBN, we shall see that τ_s is much smaller than τ_m. Another difference is that in our case, only one compartment, the "dendrite," is excitable, while the "soma" is passive as it merely introduces a second timescale [48]. Finally, in the two-compartment model, there is a conductance connecting dendrite and soma compartments g_c, while in the MSBN, this conductance is taken to be infinite; namely, the dendrite and soma are assumed to be directly connected. This is known as electrotonic coupling [39].

An important point to make is that, as mentioned in the previous section, the spiking frequency is a function of C_m, which allows for a very simple control of the timescales of neuronal behaviors. In **Figure 8(d)**, we demonstrate how the spiking traces of the MSBN can straightforwardly be tuned to biological realistic timescales by modulating C_m. This feature is very attractive for potential applications in brain-machine-interfaces and neuroprosthetics.

The MSBN is a remarkably compact circuit comprising only five conventional electronic components, yet its emergent dynamics is complex and biomimetic.

Notably, as we shall discuss later on, the neuronal behaviors generated by the MSBN are qualitatively similar to those observed in some biological neurons, such as nigral dopamine neurons under the effect of neurotoxins.

5. The phase diagram and connections to neuroscience

To better understand the origin of the various spiking states of the MSBN, we explore the phase diagram of the model. We find convenient to do this in terms of the variables input current I_{in} and the time constant of the soma compartment τ_s (C_s fixed and τ_s varied by changing R_s). In the limit of τ_s small, we recover the basic MSN that we discussed before, which shows only tonic spiking (see **Figures 6** and 7).

Figure 9 illustrates the full phase diagram. The red, blue, green, and yellow regions represent the location of the four different neuronal behaviors shown above. The gray region stands for the quiescent state, which appears when I_{in} is too small or too big with respect to the onsets of excitability (I_{in}^{min} and I_{in}^{max}), as discussed in Section 3. Examples of the spiking and bursting traces are shown on the right panels, with their respective locations indicated in the phase diagram. All four colored regions in the phase diagram are relatively large, so each neuronal behavior may be easily realized, without the need of any fine tuning.

The phase diagram gives us further insights on the mechanism operating in the MSBN neuron model. For example, on the left part of the phase diagram, where τ_s is small, the MSBN only shows Tonic Spiking, as it approaches the MSN model. This tells us that when the soma time-constant is relatively small, its impact on the dendrite time constant ($\tau_m \approx R_{gate}C_m$) is minimal. In other words, the effect of the "small"

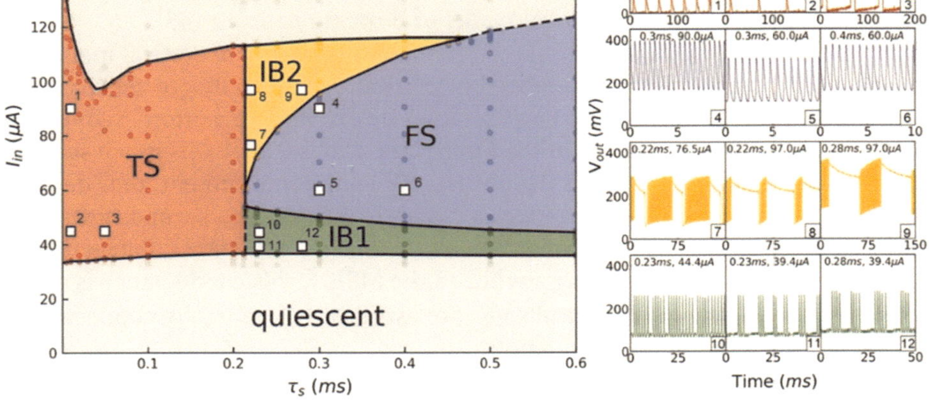

Figure 9.
Left: Phase diagram of the Memristive spiking bursting neuron circuit, with the two variables being the input current and the time-constant of the soma-compartment τ_s (C_s is fixed at 0.1 μF and τ_s varies by changing R_s). The four regions correspond to the 4 spiking types: Tonic spiking (TS), fast spiking (FS), intrinsic bursting type 1 (IB1), and intrinsic bursting type 2 (IB2). Right: Measured traces in selected locations of the phase diagram. They illustrate the effects produced by changing the I_{in} (first column) or τ_s (third column) with respect of the states shown in the second column. The corresponding positions of all traces in the phase diagram are marked with numbered white squares, with (τ_s, I_{in}) indicated in each numbered panel [reproduced from [33]].

soma can be considered as a small perturbation to the large (excitable) dendrite body, without affecting the generation of tonic spiking [48].

As τ_s gradually increases, complex neuronal behaviors emerge. Interestingly, during the phase transition from TS to other spiking states, a distinct vertical boundary appears at a certain value of τ_s. This implies that the transition from TS to other behaviors is primarily controlled by τ_s and rather independent of I_{in}. In other words, the emergences of other states occur when τ_s can no longer be considered a negligible perturbation.

We consider the transition from TS to FS as an example to describe the mechanism by which other spiking states emerge. Let us for the moment disregard the influence of C_s, and let us assume that R_s is relatively large with $R_M^{low} < < R_s < < R_M^{high}$. Initially, we start injecting current into the MSBN, with the M-device in a high-resistance (low conductance) state. Then, the membrane voltage V_m mostly drops across the M-device as R_M^{high} and R_s form a voltage divisor. As C_m becomes gradually charged, the membrane voltage V_m reaches the threshold V_{th} and provokes the M-device switching to the low-resistance state R_M^{low}. This would produce a large current from the discharge of C_m, which is the spike generation mechanism. However, and in contrast to the normal spike production mechanism in the basic MSN, now R_s is taken relatively large, $R_s > > R_M^{low}$; we have a different situation. In fact, as soon as the M-device switches to low resistance, the membrane voltage on the M-device gets dramatically reduced to close to zero. This is because the voltage divisor has drastically changed, and all the V_m now drops R_s. This has severe consequence on the stability of the spike emission, because the negligible voltage on the M-device would immediately switch it back to the high-resistance state, aborting the spike emission. As soon as the M-device switches back, it finds itself again above V_{th}, so it should switch back to low resistance. Hence, the emission mechanism now became an unstable state due to the increase of R_s. It is not hard to see that this instability may be cured by introducing a finite C_s. The (short) time-constant $\tau_s = R_s C_s$, $\tau_s \ll \tau_m$ allows a "portion" of the spike to be emitted during the short interval τ_s, during which the R_s is effectively shorted by C_s. Hence, we shall see the fast spiking behavior emerge, with the FS frequency $\sim 1/\tau_s$.

The understanding of the mechanism of IB1 and IB2 is not so straightforward. We have provided a mathematical description of its dynamics elsewhere [49]. Nevertheless, we can qualitatively see that these behaviors are associated to the upper and lower current excitability thresholds, which correspond to I_{in}^{min} and I_{in}^{max}. Hence, we can qualitatively understand that these complex intermittent spiking modes are emerging from the interaction of *two* instabilities that are in competition, namely, (i) large-R_s/low-I_{in} (IB1) and (ii) large-R_s/high-I_{in} (IB2), which also explain the "triangular" shape of those two regions of the phase diagram (green and yellow).

While our MSBN is not a bio-realistic model, it is a striking observation that it is biomimetic. Indeed, we discovered surprising similarities between the spiking modes generated by our circuit and those measured in traces of biological bursting neurons under the action of different neurotoxins that affect their normal spiking behavior. This is shown in **Figure 10**, where we directly compare MSBN traces to those from biological bursting neurons. In panel (a), we compare the different traces obtained in nigral dopamine neurons, which start from the normal (control) state and then become severely affected by exposure to neurotoxins: first apamin and then apamin +TTX + TEA [38]. These neurotoxins are found to provoke two different sets of pathological spiking behaviors, as shown in the **Figure 8(a)**. Interestingly, all the

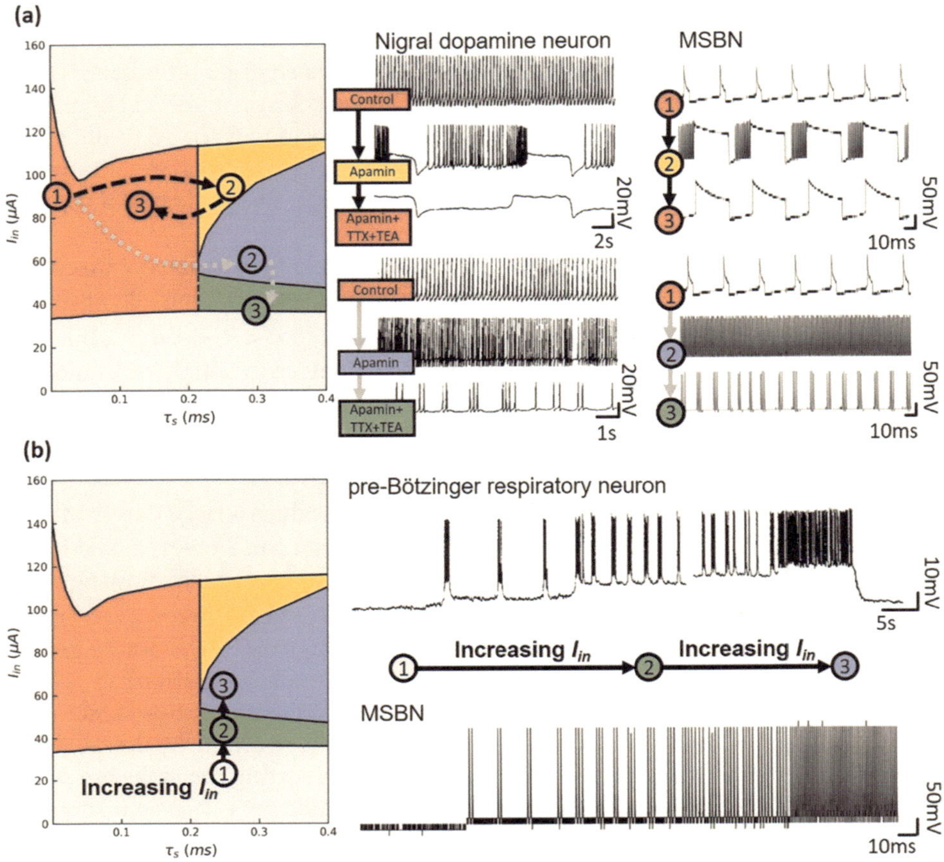

Figure 10.
The evolutionary paths in phase diagram can reproduce to some extent the evolution of biological neuronal behavior. (a) the central panel shows the successive firing states due to effects of neurotoxins on nigral dopamine neurons [adapted from [50]]. Starting from the same initial (control) tonic spiking state, two different evolutions were observed by applying the same neurotoxins. Right panel: MSBN spiking traces that are in qualitative agreement with the experimental ones. Their numbers indicated their location on the phase diagram. Left panel: The black-dashed and the gray-dotted lines show the two paths on the phase diagram, which connect the locations of the MSBN spiking traces of the right panel. (b) Neuronal behavior evolution based on the increase of I_{in}. The pre-Bötzinger respiratory neuron's behaviors are shown in the top [adapted from [51]], while the corresponding MSBN's behaviors are shown in the bottom. Measurement settings for the MSBN: in panel (a), the parameters (I_{in}, R_s) of each example trace are red 1 trace = (89.6 μA, 47 Ω), yellow 2 trace = (89.6 μA, 2.6 kΩ), red 3 trace = (82.8 μA, 1.3 kΩ), blue 2 trace = (58.8 μA, 3.0 kΩ), and green 3 trace = (36.5 μA, 3.0 kΩ); in panel (b), R_s is fixed at 2.5 kΩ, and I_{in} are increased from 26.0 μA to 36.4 μA, 40.8 μA, 46.5 μA, until 52.6 μA [reproduced from [33]].

different spiking behaviors seen in the nigral dopamine neurons can be qualitatively found in our phase diagram, as we show in the figure. It is a very interesting observation that we may then associate a "path" in the phase diagram to the action of neurotoxins. This may have a potentially useful application as a guide for neurologists to understand the origin of pathological spiking behaviors.

Perhaps even more striking is our second example shown in panel (b). In this case, the biological neurons are another type of bursters: pre-Bötzinger respiratory neurons [51]. In this case, the experiment consisted in rendering the neuron quiescent by application of a neurotoxin. Then, by sole application of input current, the neuron can be reactivated. Upon increasing input current in small steps, it is observed that the neuron emits an increasing number of short spike trains. Eventually, at high enough input current, the trains become a continuous fast spiking. Quite strikingly, we find that this same behavior can be captured by our MSBN. In fact, as shown in the **Figure 10(b)**, we observe how by moving vertically in the phase diagram, that is, solely increasing the input current, the MSBN evolves from the quiescent state through a series of bursting states with increasing number of spike trains, to a final fast spiking, in excellent qualitative agreement with the behavior of the biological pre-Bötzinger neuron [51].

6 Conclusion

In this work, we described in detail a novel volatile memristive electronic component, the M-device, which we synthetize out of two conventional out-of-the-shelf electronic components (a thyristor and a resistor).

The M-device exhibits pinched hysteresis and bears a striking qualitative similarity to the I-V behavior of many Mott insulators, which are quantum materials emerging as candidates for future neurocomputing hardware [20, 21, 24, 26, 37]. Like Mott neurons, our device has extreme simplicity enabling straightforward biomimetic spiking neuron behavior. Unlike Mott neurons, which are still hard to fabricate and function in a consistent and reliable manner, our M-device is extremely reliable and affordable, allowing for very cost-effective implementations. While neurons made out of discrete components present an evident limit to the practical number of neurons that one may envision in a neural network (up to hundreds), they provide a flexibility in circuit design that is unachievable in CMOS VLSI systems, which require long design times and fabrication. In any case, the M-device can be straightforwardly ported to CMOS VLSI technology [39] while carrying along the conceptual simplicity provided by the memristive device.

In recent years, Neuromorphic technology has become a significant application arena for volatile memristors [20, 21, 24, 26, 37]. Based on the M-device, we demonstrated here a simple Memristive Spiking Neuron (MSN) circuit, which provides a physical hardware implementation of the functionality of the popular Leaky Integrate-and-Fire (LIF) model, in fully analogic electronic. This is a concrete realization of the concept "*the circuit is the model*." Furthermore, we also demonstrated how this concept can be exploited following the insights from theoretical neuroscience models that indicate that a simple spiking neuron model requires an additional timescale to produce bursting behavior. That was directly implemented in the second circuit model, the Memristive Spiking Bursting Neuron (MSBN), which we introduced and analyzed here. Our simple 4-components MSBN was found to exhibit spiking and bursting behaviors that bear striking similarities to biological bursting neurons.

This simplicity enabled us the full exploration of the phase diagram of neuronal behavior of the MSBN. We made explicit contact with two notable examples of biological bursting neurons: Nigral Dopamine and pre-Bötzinger respiratory. Specifically, we were able to "follow the path" in the phase diagram of their pathological spiking evolution under the action of neurotoxins. These amazing connections may open up the way for potential unprecedented novel approaches to understanding pathological states such as those in Parkinson's and epilepsy, or even new ways for treating neurodegenerative diseases, though implementations in the field of neuroprosthetics and neuro-implants.

Acknowledgements

We acknowledge support from the French ANR "MoMA" project ANR-19-CE30-0020.

Conflict of interest

The authors declare no conflict of interest.

Author details

Jiaming Wu and Marcelo Rozenberg*
CNRS Laboratoire de Physique des Solides, Université Paris-Saclay, Orsay, France

*Address all correspondence to: marcelo.rozenberg@universite-paris-saclay.fr

References

[1] Chua L. Memristor-the missing circuit element. IEEE Transactions on Circuit Theory. 1971;**18**(5):507-519. DOI: 10.1109/TCT.1971.1083337

[2] Chua LO, Kang SM. Memristive devices and systems. Proceedings of the IEEE. 1976;**64**(2):209-223. DOI: 10.1109/PROC.1976.10092

[3] Di Ventra M, Pershin YV, Chua LO. Circuit elements with memory: Memristors, memcapacitors, and meminductors. Proceedings of the IEEE. 2009;**97**(10):1717-1724

[4] Jeong DS, Kim KM, Kim S, Choi BJ, Hwang CS. Memristors for energy-efficient new computing paradigms. Advanced Electronic Materials. 2016; **2**(9):1600090

[5] Zahoor F, Azni Zulkifli TZ, Khanday FA. Resistive random access memory (RRAM): An overview of materials, switching mechanism, performance, multilevel cell (MLC) storage, modeling, and applications. Nanoscale Research Letters. 2020;**15**:1-26

[6] Golonzka O, et al. Non-Volatile RRAM embedded into 22FFL FinFET technology. In: 2019 Symposium on VLSI Technology. 2019. pp. T230-T231

[7] Mehonic A, Sebastian A, Rajendran B, et al. Memristors—From in-memory computing, deep learning acceleration, and spiking neural networks to the future of neuromorphic and bio-inspired computing. Advanced Intelligent Systems. 2020;**2**(11):2000085

[8] Lanza M, Sebastian A, Lu WD, Le Gallo M, Chang MF, Akinwande D, et al. Memristive technologies for data storage, computation, encryption, and radio-frequency communication. Science. 2022;**376**(6597):eabj9979

[9] Merolla PA, Arthur JV, Alvarez-Icaza R, Cassidy AS, Sawada J, Akopyan F, et al. A million spiking-neuron integrated circuit with a scalable communication network and interface. Science. 2014;**345**(6197):668-673

[10] Marković D, Mizrahi A, Querlioz D, et al. Physics for neuromorphic computing. Nature Reviews Physics. 2020;**2**(9):499-510

[11] Thakur CS, Molin JL, Cauwenberghs G, Indiveri G, Kumar K, Qiao N, et al. Large-scale neuromorphic spiking array processors: A quest to mimic the brain. Frontiers in Neuroscience. 2018;**12**:891

[12] Mead C. Neuromorphic electronic systems. Proceedings of the IEEE. 1990; **78**(10):1629-1636

[13] Ham D, Park H, Hwang S, et al. Neuromorphic electronics based on copying and pasting the brain. Nature Electronics. 2021;**4**:635-644. DOI: 10.1038/s41928-021-00646-1

[14] Herz AVM, Gollisch T, Machens CK, et al. Modeling single-neuron dynamics and computations: A balance of detail and abstraction. Science. 2006; **314**(5796):80-85

[15] Gerstner W, Kistler WM, Naud R, et al. Neuronal Dynamics: From Single Neurons to Networks and Models of Cognition. Cambridge: Cambridge University Press; 2014

[16] Rahimi Azghadi M, Chen YC, Eshraghian JK, Chen J, Lin CY, Amirsoleimani A, et al. Complementary

metal-oxide semiconductor and memristive hardware for neuromorphic computing. Advanced Intelligent Systems. 2020;**2**(5):1900189

[17] Kim H, Mahmoodi MR, Nili H, Strukov DB. 4K-memristor analog-grade passive crossbar circuit. Nature Communications. 2021;**12**(1):5198

[18] Kim SG, Han JS, Kim H, Kim SY, Jang HW. Recent advances in memristive materials for artificial synapses. Advanced Materials Technologies. 2018;**3**(12):1800457

[19] Chang YF, Fowler B, Chen YC, Zhou F, Pan CH, Chang TC, et al. Demonstration of synaptic behaviors and resistive switching characterizations by proton exchange reactions in silicon oxide. Scientific Reports. 2016;**6**(1): 21268

[20] del Valle J, Ramírez JG, Rozenberg MJ, Schuller IK. Challenges in materials and devices for resistive-switching-based neuromorphic computing. Journal of Applied Physics. 2018;**124**(21):211101. DOI: 10.1063/ 1.5047800

[21] Yi W, Tsang KK, Lam SK, Bai X, Crowell JA, Flores EA. Biological plausibility and stochasticity in scalable VO2 active memristor neurons. Nature Communications. 2018;**9**(1):4661

[22] Park SO, Jeong H, Park J, Bae J, Choi S. Experimental demonstration of highly reliable dynamic memristor for artificial neuron and neuromorphic computing. Nature Communications. 2022;**13**(1):2888

[23] Rocco R, del Valle J, Navarro H, Salev P, Schuller IK, Rozenberg M. Exponential escape rate of filamentary incubation in Mott spiking neurons. Physical Review Applied. 2022;**17**(2): 024028

[24] Hoffmann A et al. Quantum materials for energy-efficient neuromorphic computing: Opportunities and challenges. APL Materials. 2022; **10**(7):070904

[25] Strukov D, Snider G, Stewart D, et al. The missing memristor found. Nature. 2008;**453**:80-83. DOI: 10.1038/ nature06932

[26] Pickett MD, Medeiros-Ribeiro G, Williams RS. A scalable neuristor built with Mott memristors. Nature Materials. 2013;**12**(2):114-117

[27] Del Valle J et al. Subthreshold firing in Mott nanodevices. Nature. 2019; **569**(7756):388-392

[28] Del Valle J et al. Spatiotemporal characterization of the field-induced insulator-to-metal transition. Science. 2021;**373**(6557):907-911

[29] Adda C et al. Direct observation of the electrically triggered insulator-metal transition in V 3 O 5 far below the transition temperature. Physical Review X. 2022;**12**(1):011025

[30] Li Y, Wang Z, Midya R, Xia Q, Yang JJ. Review of memristor devices in neuromorphic computing: Materials sciences and device challenges. Journal of Physics D: Applied Physics. 2018; **51**(50):503002

[31] Stoliar P et al. A leaky-integrate-and-fire neuron analog realized with a Mott insulator. Advanced Functional Materials. 2017;**27**(11):1604740

[32] Georges A, Kotliar G, Krauth W, Rozenberg MJ. Dynamical mean-field theory of strongly correlated fermion systems and the limit of infinite

dimensions. Reviews of Modern Physics. 1996;**68**(1):13

[33] Wu J, Wang K, Schneegans O, Stoliar P, Rozenberg M. Bursting dynamics in a spiking neuron with a memristive voltage-gated channel. Neuromorphic Computing and Engineering. 2023;**3**(4):044008

[34] Rozenberg MJ, Schneegans O, Stoliar P. An ultra-compact leaky-integrate-and-fire model for building spiking neural networks. Scientific Reports. 2019;**9**(1):11123

[35] Stoliar P, Schneegans O, Rozenberg MJ. Biologically relevant dynamical behaviors realized in an ultra-compact neuron model. Frontiers in Neuroscience. 2020;**14**:421

[36] Stoliar P et al. Universal electric-field-driven resistive transition in narrow-gap Mott insulators. Advanced Materials. 2013;**25**(23):3222-3226

[37] Janod E, Tranchant J, Corraze B, Querré M, Stoliar P, Rozenberg M, et al. Resistive switching in Mott insulators and correlated systems. Advanced Functional Materials. 2015;**25**(40): 6287-6305

[38] Rubin JE, Terman D. High frequency stimulation of the subthalamic nucleus eliminates pathological thalamic rhythmicity in a computational model. Journal of Computational Neuroscience. 2004;**16**:211-235

[39] Stoliar P, Akita I, Schneegans O, Hoki M, Rozenberg MJ. A spiking neuron implemented in VLSI. Journal of Physics Communications. 2022;**6**(2): 021001

[40] Sze SM, Li Y, Ng KK. Physics of Semiconductor Devices. New Jersey: John Wiley & Sons; 2021

[41] Chua L. If it's pinched it'sa memristor. Semiconductor Science and Technology. 2014;**29**(10):104001

[42] Kandel ER, Schwartz JH, Jessell TM, Siegelbaum S, Hudspeth AJ, Mack S, editors. Principles of Neural Science. Vol. 4. New York: McGraw-Hill; 2000. pp. 1227-1246

[43] Hodgkin AL, Huxley AF. A quantitative description of membrane current and its application to conduction and excitation in nerve. The Journal of Physiology. 1952;**117**(4):500

[44] Morris C, Lecar H. Voltage oscillations in the barnacle giant muscle fiber. Biophysical Journal. 1981;**35**(1): 193-213

[45] Ermentrout B, Terman DH. Mathematical foundations of neuroscience. New York: Springer; 2010

[46] Bianchi D, Marasco A, Limongiello A, Marchetti C, Marie H, Tirozzi B, et al. On the mechanisms underlying the depolarization block in the spiking dynamics of CA1 pyramidal neurons. Journal of Computational Neuroscience. 2012;**33**:207-225

[47] Pinsky PF, Rinzel J. Intrinsic and network rhythmogenesis in a reduced Traub model for CA3 neurons. Journal of Computational Neuroscience. 1994;**1**: 39-60

[48] Segev I, Rall W. Excitable dendrites and spines: Earlier theoretical insights elucidate recent direct observations. Trends in Neurosciences. 1998;**21**(11): 453-460

[49] Fernandez LE, Carpio A, Wu J, Boccaletti S, Rozenberg M, Mindlin GB. A model for an electronic spiking neuron built with a memristive voltage-gated

element. Chaos, Solitons and Fractals. 2024;**180**:114555

[50] Ping HX, Shepard PD. Apamin-sensitive Ca (2+)-activated K+ channels regulate pacemaker activity in nigral dopamine neurons. Neuroreport. 1996; **7**(3):809-814

[51] Butera RJ Jr, Rinzel J, Smith JC. Models of respiratory rhythm generation in the pre-Botzinger complex. I. Bursting pacemaker neurons. Journal of Neurophysiology. 1999;**82**(1):382-397

Section 2

Devices

Chapter 3

Oxide-Based Resistive RAM Analog Synaptic Behavior Assessment for Neuromemristive Systems

Hassen Aziza

Abstract

Resistive RAM (RRAM) synaptic plasticity behavior is essential to enable future power-efficient brain-inspired accelerators. In oxide-based RAM (so-called OxRAM), synaptic plasticity is reflected in the OxRAM conductance modulation capability, making the technology well suited for neuromorphic applications. This chapter assesses the conductance modulation of HfO_2-based RRAM and leverages experimental measurement results to establish the inherent synaptic behavior of the technology. The OxRAM conductance modulation capability is demonstrated based on a 1T1R elementary memory array test chip. The memory array OxRAM cells are operated using two distinct programming schemes. The first one is based on RESET (RST) peak voltage modulation. The second one is based on SET compliance current modulation. Both methods show that OxRAM conductance modulation is highly impacted by the variability of the technology. Hence, variability happens to be one the most challenging limitations for OxRAMs used in neuromorphic applications.

Keywords: RRAM, OxRAM, conductance modulation, synaptic plasticity, neuromorphic computing

1. Introduction

The significant growth in portable and connected devices constantly produces a huge volume of data [1]. At the same time, edge-Artificial Intelligence (AI) applications, which require processing and storage of data in close vicinity [2], accelerate the demand for more memory space. In the last 20 years, different memory concepts have been explored, aiming to mitigate the major limitations of existing solid-state memories: the volatility of RAM technology and the high latency of Flash [3]. In this context, Resistive RAM (RRAM), with its simple structure, low latency ($\sim$10 ns), small size (<10 nm), compatibility with current CMOS manufacturing processes, and low power operation, is believed to be the best contender [4].

At the system level, conventional computing technology based on the Von Neumann (VN) architecture is facing fundamental limits such as low energy efficiency and latency [5]. It results in the development of alternative information processing architectures categorized as non-Von Neumann (non-VN) [6]. To date, there has been a lot of research activities dedicated to the development of bio-inspired computing architectures such as Artificial Neuromorphic Networks (ANNs) [7]. Indeed, ANNs have demonstrated better performances in identification tasks compared to VN hardware architectures. ANNs are meant to mimic biological neuromorphic networks. They are based on units, referred to as neurons, that are connected in specific patterns. A synapse, associated with a weight, acts as a "programmable" link between each neuron. Weights generally stimulate or attenuate the signal passing through the network [8].

RRAM devices [9, 10] are seen as the optimal candidates to mimic biological synapses as they possess the capability to modulate their conductance [11, 12]. Among all the alternative RRAM technologies, OxRAMs have proven to be very effective in implementing some aspects of synaptic plasticity [13, 14]. Synaptic plasticity in OxRAMs relies on the MultiLevel Cell (MLC) storage capability of the technology, which has been reported in many studies related to neuromorphic applications [15–17]. During a SET programming operation, MLC can be applied to the OxRAM by modulating its compliance current. During an RST operation, MLC can be obtained by varying the RST voltage level. Another technique consisting in varying the pulse widths and amplitudes during SET or RST operations can also be used. However, in most prior works, MLC operation has been implemented at the device level, and MLC implementation at the circuit and system levels remains to be explored [18–20]. Also, and more importantly, RRAM variability poses a significant hurdle for RRAM widespread adoption in neuromorphic computing systems. Variability refers to the inherent differences in the electrical characteristics of RRAM devices, even when fabricated under the same manufacturing conditions. This variability can manifest as variations in key parameters of RRAM devices including the resistance levels, thus preventing reliable and predictable device behavior, which is crucial for accurate neural network operations [21].

In this chapter, we present how OxRAM memory devices can be used for synaptic plasticity functions. The conductance modulation of the OxRAM cell is demonstrated versus variability at a memory array test chip level by modulating the peak voltage applied to the cell during an RST operation or by modulating the current passing through the cell during a SET operation. This chapter is organized as follows. Section II presents the OxRAM technology along with memristor-based neuromorphic networks' basic concepts. In Section III, the conductance modulation capability of the OxRAM technology is evaluated experimentally versus variability using a dedicated test chip. In Section IV, the conductance modulation capability of OxRAM is discussed. Finally, Section V concludes the paper.

2. Background: RRAM technology for neuromemristive systems

2.1 OxRAM technology

Figure 1a shows a 1T1R OxRAM cell with one MOS transistor connected serially with the resistive element. The resistive element, presented in **Figure 1b**, is integrated in the back end of line (BEOL) of a 130 nm conventional CMOS technology, between the two last metal levels (Metal 4 and Metal 5) [22]. Layers composing the resistive element are deposited on the substrate surface by means of physical vapor deposition where a 10 nm Hafnium dioxide (HfO_2) thin film is placed on top of a TiN Bottom electrode (BE). A Ti/TiN stack is then added to form the top electrode (TE). A typical

DOI: http://dx.doi.org/10.5772/intechopen.1002782

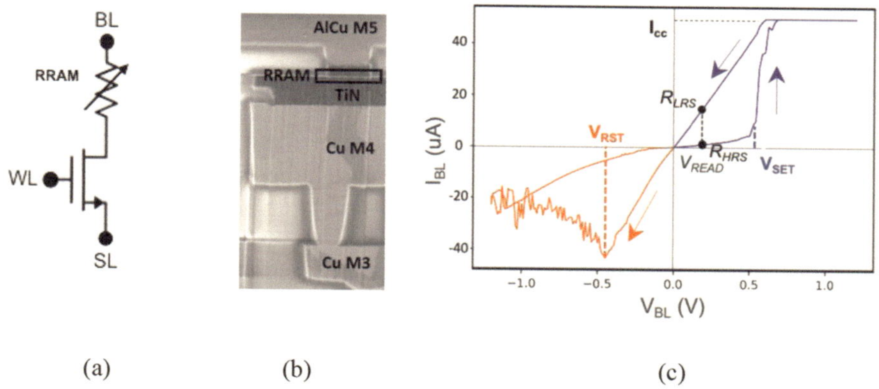

Figure 1.
(a) Symbol view of a 1T1R RRAM cell. (b) SEM cross section of the RRAM cell [20] and (c) OxRAM I–V hysteresis.

	FMG	RST	SET	Read
WL	2 V	2.5 V	2 V	2.5 V
BL	3.3 V	0 V	1.2 V	0.1 V
SL	0 V	1.2 V	0 V	0 V
$R_{LRS/HRS}$	10 kΩ	240 kΩ	15 kΩ	—

Table 1.
OxRAM cell standard operating voltages.

1T1R I-V hysteresis is presented in **Figure 1c**. According to this hysteresis, the OxRAM can be operated as follows: after a preliminary forming (FMG) step [23], the memory cell can be switched reversibly between two resistance states, the Low Resistance State (LRS) and the High Resistance State (HRS). The switching mechanism is characterized by a sudden change in the cell current. The current change is triggered by biasing the 1T1R cell with specific voltages, V_{SET} to switch from HRS to LRS during a SET operation and V_{RST} to switch from LRS to HRS during an RST operation.

FMG, RST, SET, and read bias voltages are given in **Table 1**, with the corresponding nominal resistance levels. The read operation, used to sense the memory cell state, can be performed non-destructively. Typically, a small read voltage (around 0.1 V) is used to not disturb the current state of the cell. This allows for read operations to be performed frequently without affecting the reliability of the device. R_{HRS} and R_{LRS} levels are extracted in the forward or SET direction (blue curve in **Figure 1c**, with VBL > 0). In practice, at the circuit level, the read operation is performed by measuring the current associated with the R_{LRS} and R_{HRS} resistances to determine whether the cell is in logic '0' (HRS) or the logic '1' (LRS). In the 1T1R cell architecture, the amount of current passing through the cell is controlled according to the select transistor gate voltage bias. This maximal current or clamping current is referred to as I_{CC}. From a reliability and functionality standpoint, it has been demonstrated that one of the most important limitations of OxRAM technology is related to the memory cell variability (spatial and temporal) [24, 25]. Actually, variability in R_{HRS}/R_{LRS} is so erratic that the technology has been employed as an entropy source in True Random Number Generators (TRNG) [26, 27].

2.2 OxRAM as a synapse

Synapses are central elements in biological neuromorphic networks; however, a basic electronic device emulating synapses is not available yet, thus complicating the development of integrated circuits able to mimics biological systems [28]. OxRAMs have shown the ability to be used as synapses within hybrid CMOS-memristive devices [29, 30]. Actually, the OxRAM cell has been identified as a possible candidate for building an electronic equivalent of a biological synapse as its conductance can be electrically increased or decreased [31]. In neuromorphic network architectures, OxRAM devices are meant to store synaptic weights. Furthermore, OxRAM technology can be integrated within crossbar and 3D architectures to reach integration densities never seen before [32].

Although the potential of OxRAM to mimic the synaptic behavior has been demonstrated in many studies [28], OxRAM neuromorphic network architectures also have some limitations that have been pointed out in the literature. Here are a few commonly mentioned limitations. Crossbar arrays leverage the non-linear relationship between the voltage and resistance of particular RRAM devices, which is essential to avoid the integration of the select transistor. Avoiding the use of the select transistor allows to reach the minimal $4F^2$ cell size [33]. Crossbar arrays are very difficult to implement. Indeed, without a select device, a large amount of leakage current known as sneak-path current, flowing through unselected cells, is inevitable. This phenomenon leads to a reduction of crossbar memory array sizes [34]. Concerning 3D, while crossbar prototypes including eight memory stacks have been fabricated, many of the manufacturing issues including stack-dependent variability are not fixed [35]. In addition, to be efficient, OxRAM conductance modulation requires the implementation at the circuit level of programming techniques capable of tuning accurately the analog resistance levels. On top of that, and as already mentioned, RRAM devices exhibit large variability in their electrical characteristics, affecting the accuracy and reliability of synaptic operations. Hence, addressing RRAM variability in neuromorphic computing systems has become an active area of research. Several strategies are being explored to mitigate the impact of variability, including improved fabrication processes [36], device modeling [37, 38], and circuit-level compensation approaches [39]. Additionally, machine learning algorithms that are more robust to device variations have been developed [40].

2.3 RRAM-based artificial neuromorphic networks

The fully connected neuromorphic network is the neuromemristive network topology that has been mostly explored in the literature [2, 30]. In this network, each neuron of a layer is connected to every neuron of the previous layer [11]. For clarity, a fully connected neuromorphic network made of two layers is considered in **Figure 2**. This feed-forward network can achieve classification tasks based on a linear predictor function associating a set of weights, W_{ik}, with an input vector, X_i. A vector-matrix multiplication is implemented to calculate outputs Y_k (Eq. 1 in **Figure 2**) [41]. An OxRAM cell array could naturally achieve matrix–vector multiplications in one iteration. This approach has been successfully applied experimentally in [42] using a 1T1R array of metal-oxide devices used for pattern identification. **Figure 3a** presents how the neuromorphic network can be mapped to a classical 1T1R array. Cells in a row are organized by attaching the transistor source nodes to the Source Line (SL) and attaching transistor gate nodes to the same Word Line (WL), while the cells in a

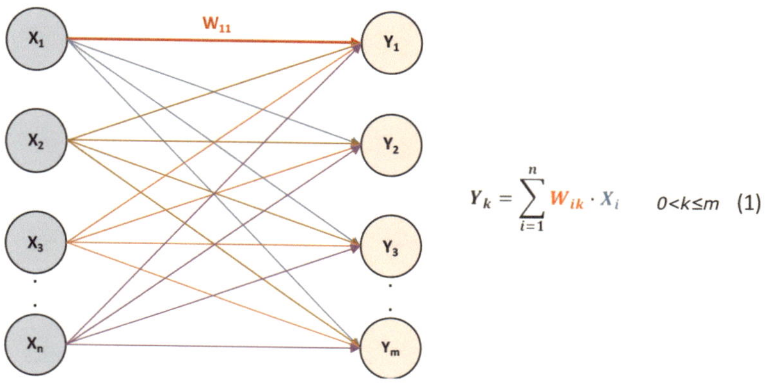

Figure 2.
Two-layer feed-forward neuromorphic network.

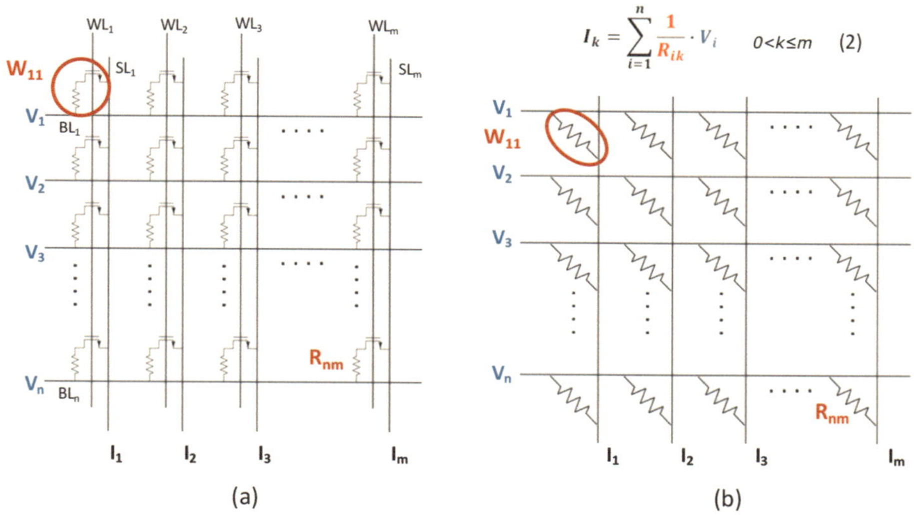

Figure 3.
(a) Matrix–vector multiplication in a 1T1R OxRAM array and (b) in 1R crossbar array.

column are arranged by attaching the top electrode of the RRAM to the bit line (BL). Input vectors (X_x) are associated with input voltages (V_x), and matrix weights (W_{ik}) are associated with memory cell resistance values R_{ik}, provided that the ON resistance of the select transistors can be neglected compared to the OxRAM HRS/LRS resistances. This construction can achieve vector-matrix multiplications according to Eq. 2 in **Figure 3**. The weighted sum is calculated by determining the total current I_k and applying an activation function to the overall current. Crossbar arrays are also considered as good candidates for implementing neuromorphic networks. The structure presented in **Figure 3b** straightforwardly implements vector-matrix multiplication without the need for select transistors. This approach has been demonstrated experimentally in [43]. However, due to the sneak-path issues, only a 12 × 12 crossbar array size was considered [44].

Figure 4a shows a multi-layer perceptron architecture including one hidden layer made of neurons. **Figure 4b** illustrates how this network can be mapped to a crossbar array. In this configuration, neurons [11] are usually designed using standard CMOS

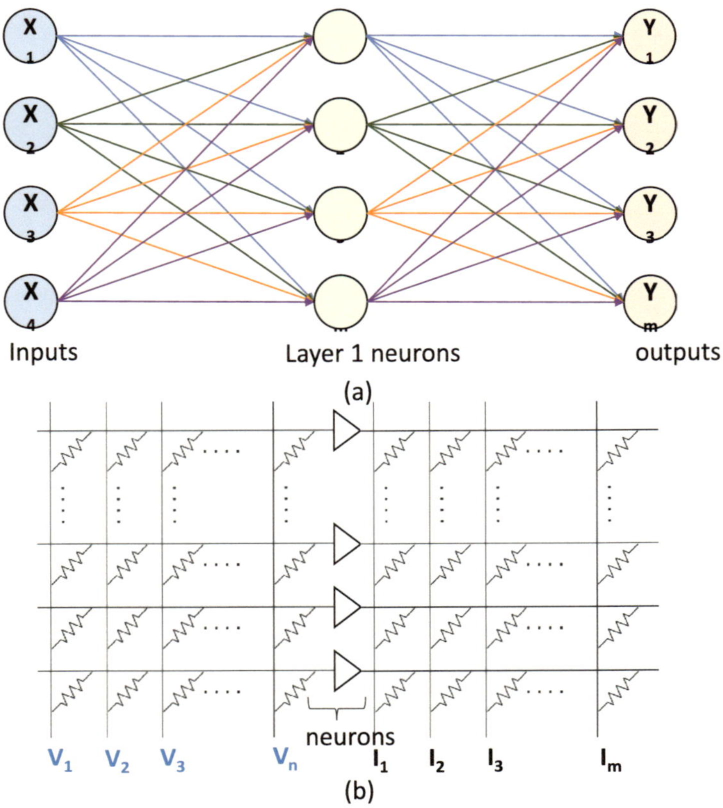

Figure 4.
(a) Multi-layer neuromorphic network with one hidden layer and (b) mapping of the multi-layer neuromorphic network into a crossbar array.

technologies. The neuron computes the weighted sum of its inputs (input vectors X_i) and applies an activation function to obtain a signal, which is conveyed to the next stage (outputs Y_k).

For each network architecture, the key point is the ability of the OxRAM device to provide multiple resistance values to store synaptic weights. Indeed, before the inference stage where vector-matrix multiplication algorithms are executed, a set of weights are loaded into the network. In this stage, all the OxRAM cells of the network are programmed to a precise conductance level with respect to the pre-defined weights. Hence, the evaluation of the conductance modulation capability of OxRAM devices in a neuromorphic network context is of the prime importance.

3. OxRAM synaptic plasticity experimental validation

To change OxRAM cells' conductance values, two different approaches are considered. The first one is based on controlling the maximum voltage during the RST operation [45], and the second one is based on compliance current modulation during the SET operation [4]. Indeed, the RST programming step is a voltage-dependent process, whereas the SET programming step is a current-dependent process [4].

Another technique consisting in changing the pulse width of RST and SET voltages is also considered [18] but is not developed in this chapter. To evaluate the two above-mentioned methods, a dedicated test chip is considered for measurements.

3.1 Test chip presentation and experimental setup

Figure 5 shows the schematic view of the test chip used for measurements. OxRAM cells are arranged to form eight 8-bit memory words. The active row is selected through the Word Line (WL_X) signal. During a SET operation, bit line (BL_X) signals are used to select specific columns. During an RST operation, signals applied to Source Lines (SL_X) are used to RST a whole word or a specific cell. BL, WL, and SL nodes are externally available in order to allow a full flexibility during characterization. Measurements are performed using a Keysight B1500 semi-conductor parameter analyzer [26]. During characterization, R_{LRS} and R_{HRS} extraction is conducted using 1 ms voltage sweeps: the applied voltage increases step by step, and the current flowing through the cell is measured. **Figure 5b** presents a layout view of the memory array. Due to the limited pin out of the probe card, a subset of the 8x8 memory array is used for our experiments (7 × 7 memory array).

Preliminary results show that the memory array electrical characteristics are affected by variability. **Figure 6** shows R_{HRS} and R_{LRS} distributions. 100 consecutive RST/SET operations applied to the memory array (49 cells × 100) are performed to obtain these distributions. Impact of the variability on HRS and LRS is clearly visible with an HRS distribution spread more marked with respect to the LRS one. R_{HRS} and R_{LRS} distributions are extracted using the following measurement protocol: before any operation, the memory array is first formed. Then, memory cells are RST one by one to extract the R_{HRS} value at 0.1 V. After RST, cells are SET to extract the R_{LRS} value, also at 0.1 V. The RST/SET process is repeated 100 times to obtain the resistance distributions. C2C and D2D variabilities are also revealed at the I–V characteristic

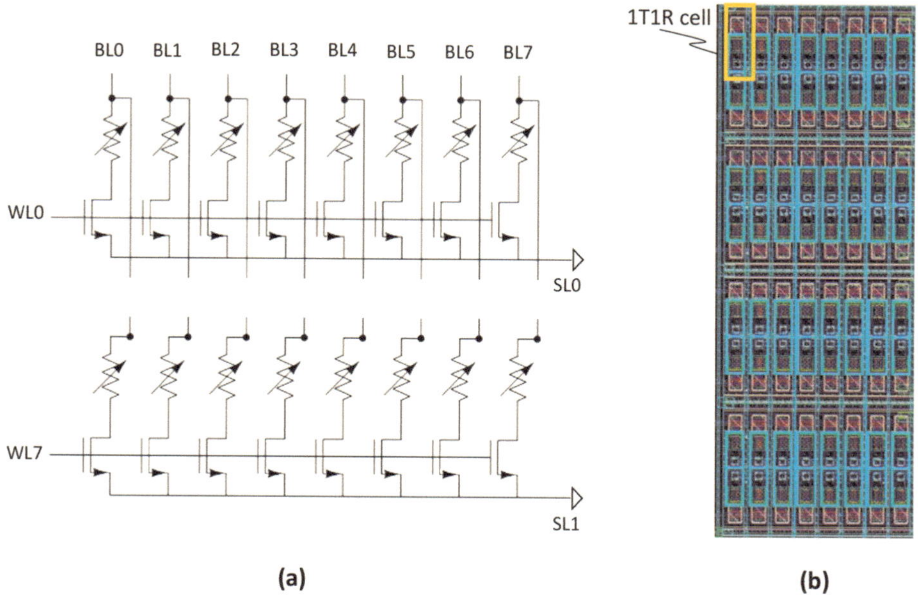

Figure 5.
(a) 8 × 8 OxRAM memory array and (b) layout of the memory array fabricated in a 130 nm technology.

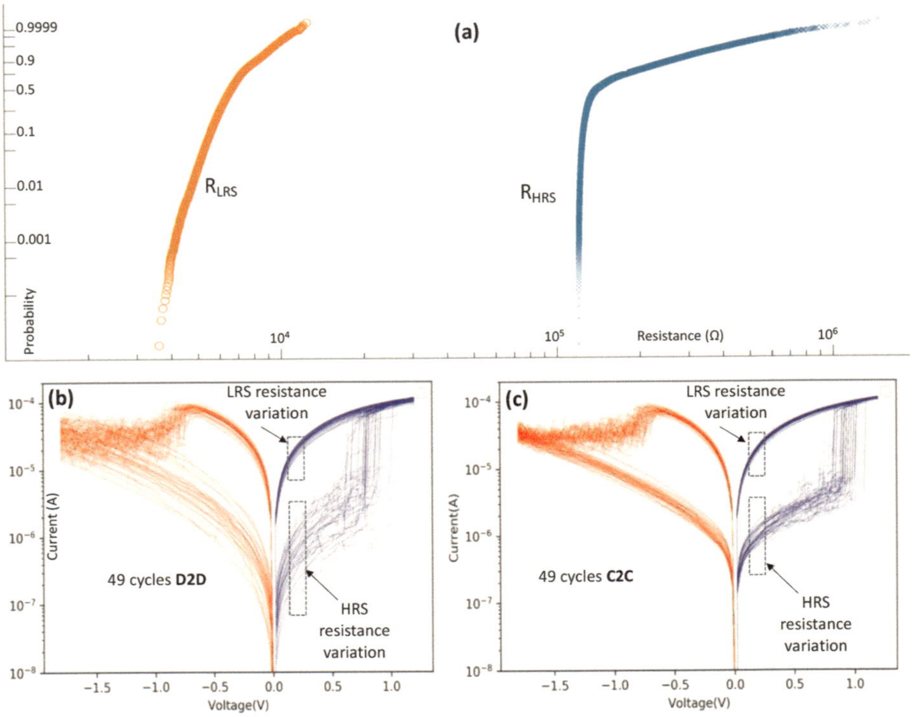

Figure 6.
(a) OxRAM HRS and LRS resistance distributions. OxRAM I–V characteristics highlighting (b) D2D and (c) C2C variabilities.

level after an RST/SET operation is applied to each of the 49 cells of the memory to catch D2D variability (**Figure 6b**) and after an RST/SET operation is applied 49 times to an isolated cell to catch C2C variability (**Figure 6c**). HRS and LRS resistance values are both affected by C2C and D2D variabilities, with a more marked impact in the HRS state, particularly for D2D variability. As a conclusion, these preliminary results show that the variability, due to the inherent nature of OxRAM devices, has to be considered when designing OxRAM-based neuromorphic networks.

3.2 RST voltage control

Different RST voltage biases can be applied to an OxRAM cell to control its resistance. Higher RST voltages result in higher resistance levels, while lower RST voltages lead to lower resistance levels. In our experiment, the conductance modulation is achieved by controlling the peak RST voltage referred to as V_{stop}. The measurement protocol is presented in **Figure 7a**. The protocol starts with the first RST operation (RST_1) at 1.8 V peak voltage in order to set the cell in a steady HRS state. Then, a SET operation at 1.2 V peak voltage is performed to switch the cell to the LRS state. A second RST operation (RST_2) with an increasing V_{stop} voltage is applied to the cell. V_{stop} ranges from 0.6 V to 1.75 V, with a 0.05 V voltage step in each iteration (24 different maximum V_{stop} peak voltages are considered). In each iteration, the HRS resistance variation is evaluated during a read operation. In **Figure 7b, I–V** curves are presented for a single cell located in the middle of the memory array. I–V characteristics are extracted following the RST_1-SET-RST_2 programming sequence with a V_{stop}

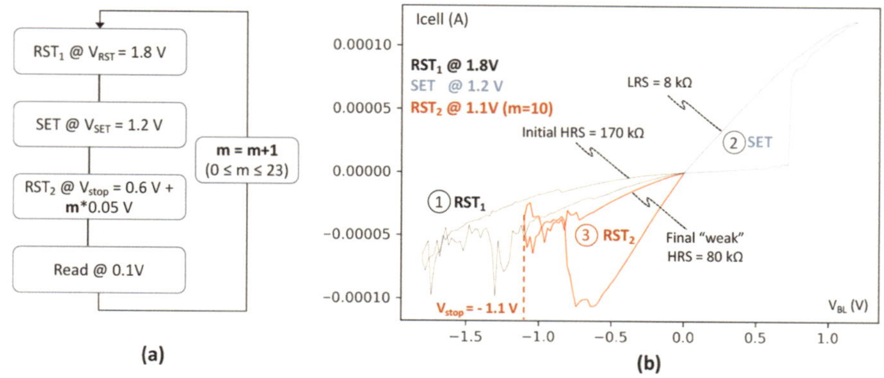

Figure 7.
(a) R_{HRS} modulation protocol and (b) resulting I–V characteristics.

value equals to 1.1 V. The initial state of the cell is a "weak" RST state before the first RST_1 programming operation. The RST_1 operation results in a contracted I-V hysteresis (black curve, numbered 1). From this initial HRS state (170 kΩ), a SET programming operation (blue curve, numbered 2) switches the cell to an LRS state (8 kΩ). Lastly, a "weak" RST operation, with V_{stop} equals to 1.1 V is applied to switch back the cell to an intermediate HRS state (red curve, numbered 3). In fact, as the RST operation is not entirely finalized due to the clamping value V_{stop}, the OxRAM cell ends up in an intermediate HRS resistance state (80 kΩ).

$R_{HRS}(V_{stop})$ plot is presented in **Figure 8** for two single cells. We can see that $R_{HRS}(V_{stop})$ is not a monotonically increasing function of V_{stop}. The curve shows some abrupt resistance changes: resistance increase and resistance drops, both highlighted with red circles in **Figure 8a** and **b**, respectively.

Figure 9a presents the evolution of $R_{HRS}(V_{stop})$ for a single cell cycled 10 times to highlight C2C variability. An increasing trend of $R_{HRS}(V_{stop})$ is observed, combined with a high variability. This increasing trend is verified in **Figure 9b**, which presents a curve resulting from an averaging performed on the 10 curves presented in **Figure 9a**. Also, for each V_{stop} value, the measurement protocol of **Figure 7a** is applied to all the cells of the memory array (49 curves) in order to catch D2D variability. An important variability is also observed (see **Figure 10a**). This variability is mitigated after

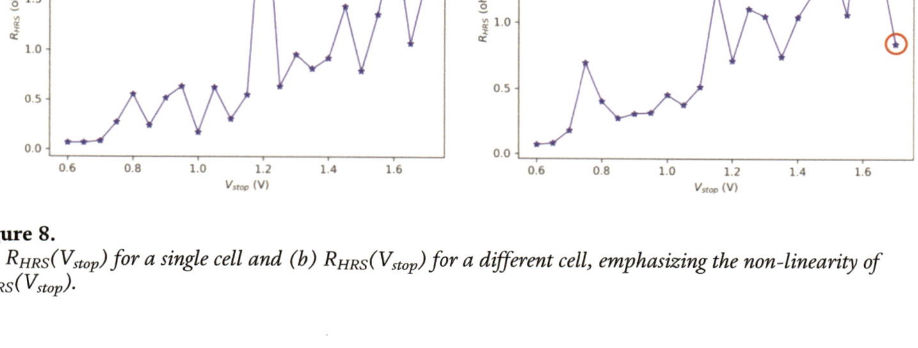

Figure 8.
(a) $R_{HRS}(V_{stop})$ for a single cell and (b) $R_{HRS}(V_{stop})$ for a different cell, emphasizing the non-linearity of $R_{HRS}(V_{stop})$.

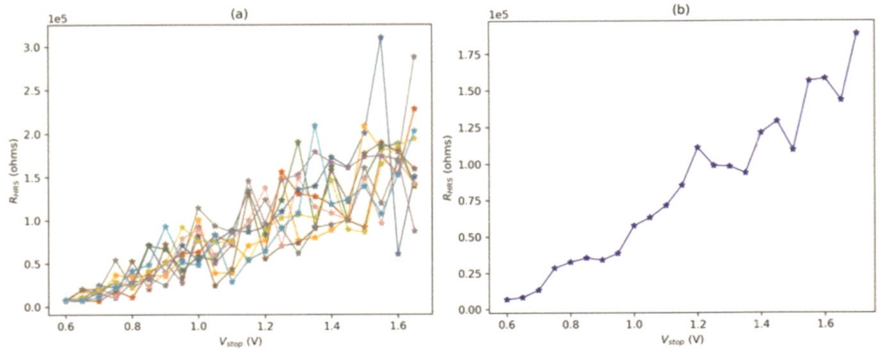

Figure 9.
(a) $R_{HRS}(V_{stop})$ for a single cell cycled 10 times (C2C variability). (b) Average curve computed from the 10 curves presented in (a).

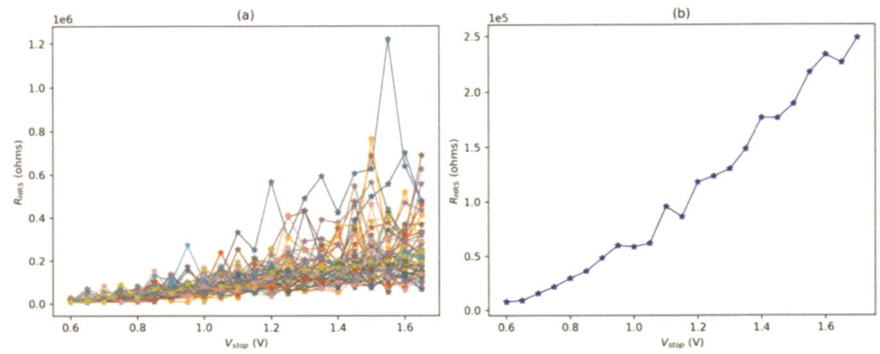

Figure 10.
(a) $R_{HRS}(V_{stop})$ for 49 single cells (D2D variability). (b) Average plot computed from the 49 curves presented in (a).

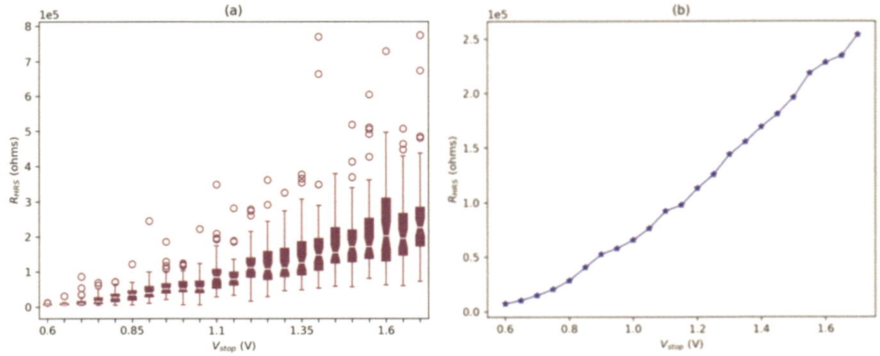

Figure 11.
(a) $R_{HRS}(V_{stop})$ box plots. Each box plot includes 49 × 10 measurements (C2C and D2D variabilities) for each V_{stop} value. (b) R_{HRS} mean value vs. V_{stop} obtained from box plots presented in (a).

averaging the 49 curves (see **Figure 10b**). Note that whether considering C2C or D2D variability, the averaging still provides a non-monotonic function of $R_{HRS}(V_{stop})$.

The C2C and D2D variability impact on R_{HRS} is observed in **Figure 11a**, with box plots related to the same memory array programmed 10 times (box plots made of

49x10 measurements for each V_{stop} values). For this latter case, **Figure 11b** shows that the smoothing effect due to the averaging alleviates the variability: the median value of $R_{HRS}(V_{stop})$ is turned monotonic, meaning that the mean value of R_{HRS} strictly increases with increasing V_{stop} values.

These experimental results clearly show that the impact of variability on memristive ANN parameters, such as accuracy, requires to be evaluated if the conductance modulation is achieved using RST voltage control.

3.3 Compliance current modulation during SET

Conductance modulation in OxRAM can also be achieved by changing the compliance current I_C, where I_C refers to the maximum current allowed to flow through the cell during the SET operation. Indeed, it has been demonstrated that the final cell resistance is a function of I_C [28]. Higher I_C values lower the LRS resistance, while lower I_C values increase the LRS resistance. In the 1T1R cell, the current involved in the SET operation is controlled by changing the gate transistor voltage V_{WL} (see **Figure 1a**). The compliance current modulation measurement protocol is presented in **Figure 12a**. First, OxRAM cells are RST to start from a nominal HRS state. After RST, a SET operation is achieved with a variable V_{WL} voltage ("weak" SET) to enter the LRS state. V_{WL} ranges from 1 V to 2 V, with a 0.1 V voltage step in each iteration, giving 11 distinct V_{WL} voltage levels. In each iteration, the LRS resistance is measured during a read operation.

Figure 12b shows the effect of three different V_{WL} voltage levels (2 V, 1.5 V, and 1 V) on a single cell. When V_{WL} decreases, the maximum SET current decreases, resulting in three distinctive LRS levels (13 kΩ, 15 kΩ, and 34 kΩ), while the HRS resistance stays in the same range with a value of around 300 kΩ. The evolution of $R_{LRS}(V_{WL})$ is presented in **Figure 13a** for increasing V_{WL} values (one single cell is considered). An increase of R_{LRS} (i.e., current slope increase) is clearly visible. **Figure 13b** presents the evolution of the LRS current versus V_{WL} while considering D2D variability. A straight impact of variability on the different LRS resistance levels is clearly visible.

Figure 14a considers four isolated cells and shows that $R_{LRS}(V_{WL})$ is not a monotonic function of V_{WL}. Yet, resistance drops are less marked compared to the RST voltage approach (**Figure 10a**). **Figure 14b** shows a box plot representation of $R_{LRS}(V_{WL})$ for all

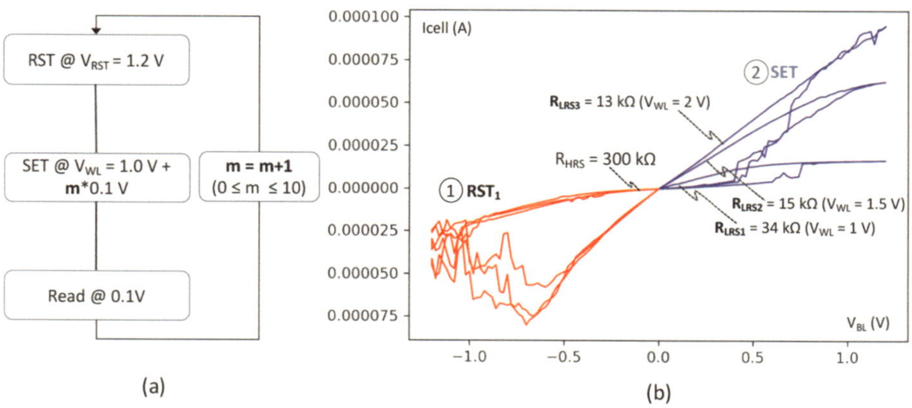

Figure 12.
(a) Compliance current modulation measurement protocol and (b) $R_{LRS}(V_{WL})$ for three different V_{WL} values (2 V, 1.5 V and 1 V).

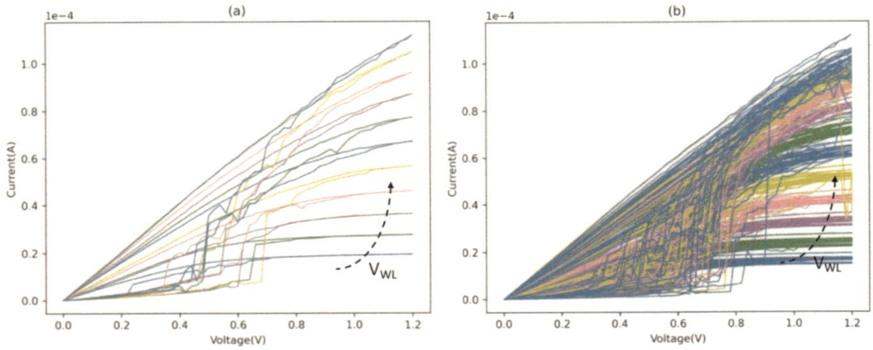

Figure 13.
(a) $R_{LRS}(V_{WL})$ for 11 V_{WL} voltage values. (b) Effect of D2D variability on $R_{LRS}(V_{WL})$ plots.

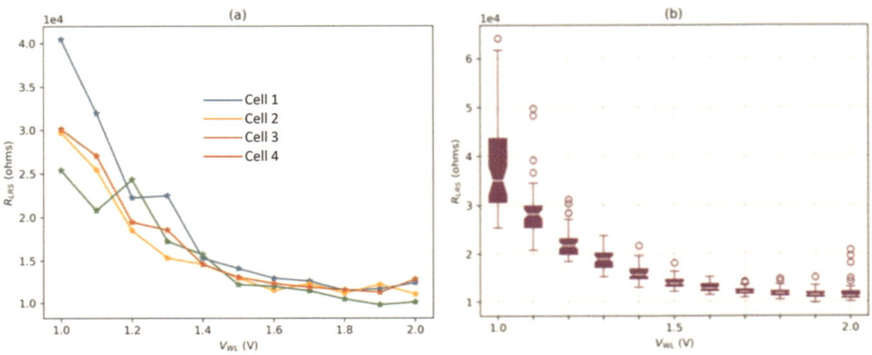

Figure 14.
(a) $R_{LRS}(V_{WL})$ for four isolated cells and (b) $R_{LRS}(V_{WL})$ box plots. Each box plot gathers 49 measurements.

the memory array cells (49 measurements for each V_{WL} level). The higher the select transistor gate voltage value, the lower the R_{LRS} after SET. Furthermore, for low V_{WL} values, R_{LRS} values are more spread. Conversely, when the compliance current I_C is high, the SET operation is more efficient, generating tighter distributions [29].

4. Discussion

In both compliance current and RST voltage control approaches, the idea was to define a relationship between the SET compliance current and the RST voltage versus R_{HRS} or R_{LRS}, respectively. R_{HRS} or R_{LRS} variations were expected to follow a monotonic evolution to confidently implement a consistent conductance modulation scheme. In this context, the proposed study shows that the variability of OxRAM technology results in resistance variations not easily compatible with the neuromorphic network architecture design.

At the cell level, the conductance modulation can be seen as a segmentation of the current-voltage plane by different I-V line segments. For simplicity, only 8 I-V characteristics are presented in **Figure 15**. Each characteristic has a slope of G_x equals to 1/ R_x, where x represents the number of resistance states ranging from 0 to n. **Figure 15** shows that a reliable conductance modulation strategy does not just depend on the programming operation but also depend on an accurate read mechanism. As already

DOI: http://dx.doi.org/10.5772/intechopen.1002782

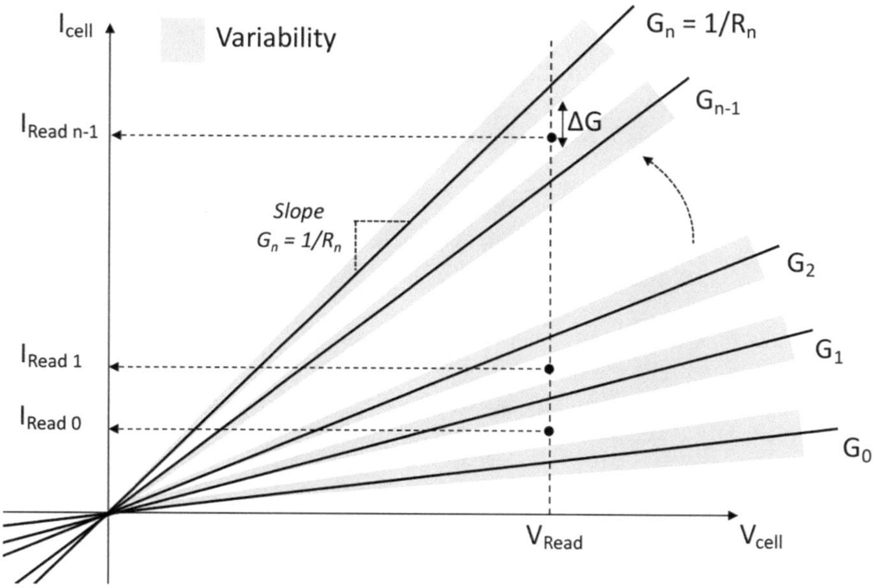

Figure 15.
I–V plane segmentation after a programming operation. Variability is illustrated by a shaded area encompassing each I–V characteristic. During a read operation, the OxRAM select transistor gate is biased (V_{Read} voltage), and different currents I_{Read} are sensed.

mentioned in the introduction, a read operation is conducted by applying a low bit line voltage bias V_{Read} to the OxRAM cell transistor and sensing the resulting currents referred to as $I_{Read\ 0}$ to $I_{Read\ n\text{-}1}$. Each current reflects the OxRAM conductance value. Several parameters, including variability, can impact the conductance margin ΔG and ultimately result in I-V characteristic overlaps. To consider the variability of the n HRS/LRS resistance states, a conductance margin, represented by shaded strips encompassing each I-V characteristic, is added in **Figure 15**.

In addition to variability, another important limitation needs to be considered for the reliable use of OxRAM in neuromorphic network architectures. This limitation is known as soft errors. A soft error is a random and non-permanent upsetting of the data stored in a memory cell, in contrast to a hard error, which is a non-recoverable failure of a cell. It has been demonstrated that OxRAM resistance can get unpredictably stuck at one resistance state for some time and suddenly recover the normal operation. **Figure 16** presents the evolution of R_{HRS} and R_{LRS} extracted at 0.1 V versus the number of programming cycles. Each programming cycle ranges from 0 to 11,269. When analyzing **Figure 16**, many HRS and LRS level overlaps are clearly visible. These overlaps happen either because one cell stops meeting its nominal characteristics or because one cell is stuck at a resistance state for one or several cycles before recovering. However, no specific pattern has been detected, demonstrating the unpredictable nature of the OxRAM technology.

These overlaps can be explained from a physical standpoint. If during a SET operation too many bonds between some of the metal and oxygen ions in the oxide layer break, too many oxygen ions are attracted to the top electrode, hence leaving oxygen vacancies acting as a conductive filament (CF). Therefore, the voltage applied across the cell to induce an effective RST operation might not be high enough. In such a case, switching the cell may demand several programming cycles in order to get those oxygen ions to drift back [46]. Conversely, if during an RST operation too many

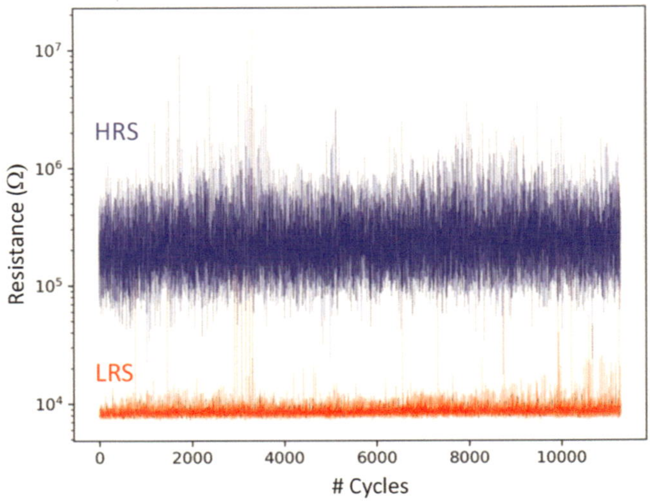

Figure 16.
R_{HRS}/R_{LRS} in log scale over 200 and 30 cycles applied to the 7 × 7 memory array.

ions are displaced, the voltage needed to induce an efficient SET operation in the next programming cycle may not be sufficient, leading the device to be stuck at HRS for several cycles [47, 48]. To mitigate soft error impact on memristive networks, it is necessary to design specific circuits able to monitor the effectiveness of each RST/SET operation [18, 49]. Also, it has a been reported at the memory array level that during consecutive RST/SET operations, write disturb can occur and impact the performance of the memory cells. The write disturb occurs when a neighboring memory cell is affected or altered during a write operation applied to a specific memory cell, leading to inadvertent resistance changes [50]. Moreover, over time, a typical bottleneck for OxRAMs lies in the endurance and retention reliability issues. These two reliability metrics narrow the resistance window, hence reducing significantly the conductance modulation capability [51, 52]. Note that cycling and endurance can lead to hard errors (memory cell stuck at one state forever). Likewise to other emerging memory technologies, OxRAM is inclined to defects [53] due to manufacturing conditions. Hence, appropriate test strategies are required to detect OxRAM-related failures [54]. However, in addition to conventional faults, OxRAM, due to its specific manufacturing process and switching mechanism, can be impacted by unique faults such as over-forming and undefined state faults [55]. Therefore, a complete fault analysis methodology needs to be defined for this memory class. Alternatively, OxRAM memristive networks have the potential to be used in harsh spatial environments [56, 57]. In such conditions, there are certain considerations and challenges that need to be addressed for OxRAM to operate reliably, such as high levels of radiations, including ionizing particles [58, 59]. These radiations can cause unwanted effects, such as resistance drifts during the OxRAM device operation [12]. To mitigate theses effects, OxRAM devices need to be radiation-hardened or shielded [12].

5. Conclusion

This study assesses the potential of OxRAMs to offer different conductance levels. The study evaluates two different programming schemes: RST voltage modulation and SET current modulation. Measurement results have established that the two programming approaches provide resistance levels highly affected by variability, which prevents an accurate control of OxRAM resistances. The RST voltage control approach shows high variability. The SET current control approach shows a more contained variability. However, in the latter approach, the variability grows exponentially with decreasing compliance currents. According to the presented experimental results, variability appears to be the main hurdle for OxRAMs used in ANN. Indeed, these systems rely on training algorithms to adjust the synaptic weights and optimize the network performances. Variability in OxRAMs can affect the convergence of these algorithms. As the device exhibits different characteristics, the learning stage might become challenging, although techniques such as adaptive learning algorithms and training strategies that account for device variability can help address this limitation. Moreover, for OxRAM technologies suffering from limited endurance, frequent updates of synaptic weights during training become even more challenging. However, ongoing research tries to address all these challenges as advanced device designs, materials engineering, circuit techniques, and system-level optimizations are being explored to improve the performance, reliability, and energy efficiency of OxRAM-based neuromemristive systems.

Author details

Hassen Aziza
M2NP, UMR CNRS 7334, Aix-Marseille Université, Marseille, France

*Address all correspondence to: hassen.aziza@univ-amu.fr

References

[1] Aziza H. Embedded memories. In: Chapter 7 in Silicon Systems for Wireless Lan 22. World Scientific; 2020. p. 199. DOI: 10.1142/9789811210723_0007

[2] Lee Y-L, Tsung P-K, Wu M. Techology trend of edge AI. In: Symposium on VLSI Design, Automation and Test (VLSI-DAT). IEEE; 2018. DOI: 10.1109/VLSI-DAT.2018.8373244

[3] Shimeng Y, Chen P-Y. Emerging memory technologies: Recent trends and prospects. IEEE Solid-State Circuits Magazine. 2016;**8**(2):43-56. DOI: 10.1109/MSSC.2016.2546199

[4] Ielmini D. Resistive switching memories based on metal oxides: Mechanisms, reliability and scaling. Semiconductor Science and Technology. 2016;**31**(6):1-25. DOI: 10.1088/0268-1242/31/6/063002

[5] Linn E et al. Logic operations in passive crossbar arrays alongside memory operations. Nanotechnology. 2012;**23**(30):305205. DOI: 10.1088/0957-4484/23/30/305205

[6] Roy D et al. Neural networks at the edge. In: 2019 IEEE International Conference on Smart Computing (SMARTCOMP). IEEE; 2019. pp. 45-50. DOI: 10.1109/SMARTCOMP.2019.00027

[7] Indiveri G et al. Neuromorphic silicon neuron circuits. Frontiers in Neuroscience. 2011;**5**:73. DOI: 10.3389/fnins.2011.00073

[8] Bi G-q, Poo M-m. Synaptic modifications in cultured hippocampal neurons: Dependence on spike timing, synaptic strength, and postsynaptic cell type. The Journal of Neuroscience. 1998; **18**(24):10 464-10 472. DOI: 10.1523/JNEUROSCI

[9] Vatajelu EI et al. Nonvolatile memories: Present and future challenges. In: Design & Test Symposium (IDT), 2014 9th International. IEEE; 2014. pp. 61-66. DOI: 10.1109/IDT.2014.7038588

[10] Bazzi H et al. RRAM-based non-volatile SRAM cell architectures for ultra-low-power applications. Analog Integrated Circuits and Signal Processing. 2021;**106**:351-361. DOI: 10.1007/s10470-020-01587-z

[11] Aziza H et al. A capacitor-less CMOS neuron circuit for neuromemristive networks. In: 2019 17th IEEE International New Circuits and Systems Conference (NEWCAS). IEEE; 2019. DOI: 10.1109/NEWCAS44328.2019.8961278

[12] Vaz PI et al. Improving TID radiation robustness of a CMOS OxRAM-based neuron circuit by using enclosed layout transistors. IEEE Transactions on Very Large Scale Integration (VLSI) Systems. 2021;**29**(6):1122-1131. DOI: 10.1109/TVLSI.2021.3067446

[13] Aziza H et al. An augmented OxRAM synapse for spiking neural network (SNN) circuits. In: 2019 14th International Conference on Design & Technology of Integrated Systems in Nanoscale Era (DTIS). IEEE; 2019. pp. 1-5. DOI: 10.1109/DTIS.2019.8735057

[14] Aziza H, Hamdioui S, Fieback M, Taouil M, Moreau M. Density enhancement of RRAMs using a RESET write termination for MLC operation. In: 2021 Design, Automation & Test in Europe Conference & Exhibition

(DATE). Grenoble, France: IEEE; 2021. pp. 1877-1880. DOI: 10.23919/DATE51398.2021.9473967

[15] Luo Y, Han X, Ye Z, Barnaby H, Seo J-S, Yu S. Array-level programming of 3-bit per cell resistive memory and its application for deep neural network inference. IEEE Transactions on Electron Devices. 2020;**67**(11):4621-4625. DOI: 10.1109/TED.2020.3015940

[16] Milo V et al. Accurate program/Verify schemes of resistive switching memory (RRAM) for In-memory neural network circuits. IEEE Transactions on Electron Devices. 2021;**68**(8):3832-3837. DOI: 10.1109/TED.2021.3089995

[17] Woo J, Moon K, Song J, Kwak M, Park J, Hwang H. Optimized programming scheme enabling linear potentiation in filamentary HfO2 RRAM synapse for neuromorphic systems. IEEE Transactions on Electron Devices. 2016;**63**(12):5064-5067. DOI: 10.1109/TED.2016.2615648

[18] Zhao L et al. Improved multi-level control of RRAM using pulse-train programming. In: IEEE International Symposium on VLSI Technology, Systems and Application (VLSI-TSA). IEEE; 2014. DOI: 10.1109/VLSI-TSA.2014.6839673

[19] Wang Y et al. Investigation of resistive switching in Cu-doped HfO2 thin film for multilevel non-volatile memory applications. Nanotechnology. 2009;**21**(4):045202. DOI: 10.1088/0957-4484/21/4/045202

[20] Freitas P, Chai Z, Zhang W, Zhang JF, Marsland J. Impact of RTN and variability on RRAM-based neural network. In: 2020 IEEE 15th International Conference on Solid-State & Integrated Circuit Technology (ICSICT). Kunming, China; 2020. pp. 1-4. DOI: 10.1109/ICSICT49897.2020.9278290

[21] Mohan C, Camuñas-Mesa LA, De La Rosa JM, Vianello E, Serrano-Gotarredona T, Linares-Barranco B. Neuromorphic low-power inference on Memristive crossbars with on-Chip offset calibration. IEEE Access. 2021;**9**:38043-38061. DOI: 10.1109/ACCESS.2021.3063437

[22] Grossi A et al. Fundamental variability limits of filament-based RRAM. In: 2016 IEEE International Electron Devices Meeting (IEDM). 2016. pp. 4.7.1-4.7.4. DOI: 10.1109/IEDM.2016.7838348

[23] Chakrabarti B et al. Multilevel switching in forming-free resistive memory devices with atomic layer deposited HfTiOx nanolaminate. IEEE Electron Device Letters. 2013;**34**(7):867-869. DOI: 10.1109/LED.2013.2262917

[24] Aziza H, Bocquet M, Portal J-M, Muller C. Evaluation of OxRAM cell variability impact on memory performances through electrical simulations. In: 2011 11th Annual Non-Volatile Memory Technology Symposium Proceeding. IEEE; 2011. pp. 1-5. DOI: 10.1109/NVMTS.2011.6137089

[25] Hajri B, Aziza H, Mansour MM, Chehab A. RRAM device models: A comparative analysis with experimental validation. IEEE Access. 2019;**7**:168963-168980. DOI: 10.1109/ACCESS.2019.2954753

[26] Aziza H et al. True random number generator integration in a resistive RAM memory Array using input current limitation. IEEE Transactions on Nanotechnology. 2020;**19**:214-222. DOI: 10.1109/TNANO.2020.2976735

[27] Postel-Pellerin J et al. True random number generation exploiting SET voltage variability in resistive RAM memory arrays. In: 2019 19th Non-Volatile Memory Technology Symposium (NVMTS). IEEE; 2019. pp. 1-5. DOI: 10.1109/NVMTS47818.2019.9043369

[28] Yu S et al. Stochastic learning in oxide binary synaptic device for neuromorphic computing. Frontiers in Neuroscience. 2013;7:186. DOI: 10.1109/ISCAS.2014.6865321

[29] Aziza H et al. Resistive RAMs as analog trimming elements. Solid-State Electronics. 2018;**142**:52-55. DOI: 10.1016/j.sse.2018.02.005

[30] Papandroulidakis G. Experimental demonstration of RRAM-based computational cells for reconfigurable mixed-signal neuro-inspired circuits and systems [Doctoral thesis]. University of Southampton; 2021

[31] Lee SR et al. Multi-level switching of triple-layered TaOx RRAM with excellent reliability for storage class memory. IEEE Symposium on VLSI Technology (VLSIT). 2012:71-72. DOI: 10.1109/VLSIT.2012.6242466

[32] Esmanhotto E et al. High-density 3D monolithically integrated multiple 1T1R multi-level-cell for neural networks. In: 2020 IEEE International Electron Devices Meeting (IEDM). San Francisco, CA, USA: IEEE; 2020. pp. 36.5.1-36.5.4. DOI: 10.1109/IEDM13553.2020.9372019

[33] Kim S, Zhou J, Lu WD. Crossbar RRAM arrays: Selector device requirements during write operation. IEEE Transactions on Electron Devices. 2014;**61**(8):2820-2826. DOI: 10.1109/TED.2014.2327514

[34] Liang J et al. Cross-point memory array without cell selectors-device characteristics and data storage pattern dependencies. TED. 2010;**57**(10):2531-2538. DOI: 10.1109/TED.2010.2062187

[35] Kim B, Li H. Leveraging 3D vertical RRAM to developing neuromorphic architecture for pattern classification. In: 2020 IEEE Computer Society Annual Symposium on VLSI (ISVLSI). Limassol, Cyprus: IEEE; 2020. pp. 258-263. DOI: 10.1109/ISVLSI49217.2020.00054

[36] Pouyan P, Amat E, Hamdioui S, Rubio A. RRAM variability and its mitigation schemes. In: 2016 26th International Workshop on Power and Timing Modeling, Optimization and Simulation (PATMOS). Bremen, Germany: IEEE; 2016. pp. 141-146. DOI: 10.1109/PATMOS.2016.7833679

[37] Baroni A et al. Low conductance state drift characterization and mitigation in resistive switching memories (RRAM) for artificial neural networks. IEEE Transactions on Device and Materials Reliability. 2022;**22**(3):340-347. DOI: 10.1109/TDMR.2022.3182133

[38] Hajri B, Mansour MM, Chehab A, Aziza H. Oxide-based RRAM models for circuit designers: A comparative analysis. In: 2017 12th International Conference on Design & Technology of Integrated Systems in Nanoscale Era (DTIS). Palma de Mallorca, Spain: IEEE; 2017. pp. 1-6. DOI: 10.1109/DTIS.2017.7930176

[39] Aziza H, Moreau M, Fieback M, Taouil M, Hamdioui S. An energy-efficient current-controlled write and read scheme for resistive RAMs (RRAMs). IEEE Access. 2020;**8**:137263-137274. DOI: 10.1109/ACCESS.2020.3011647

[40] Liu C et al. A high accuracy and robust machine learning network for

pattern recognition based on binary RRAM devices. In: 2017 International Symposium on VLSI Technology, Systems and Application (VLSI-TSA). Hsinchu, Taiwan: IEEE; 2017. pp. 1-2. DOI: 10.1109/VLSI-TSA.2017.7942472

[41] Prezioso M, Merrikh-Bayat F, Hoskins BD, Adam GC, Likharev KK, Strukov DB. Training and operation of an integrated neuromorphic network based on metal-oxide memristors. Nature. 2015;**521**(7550):61. DOI: 10.1038/nature14441

[42] Yao P et al. Face classification using electronic synapses. Nature Communications. 2017;**8**:15199. DOI: 10.1038/ncomms15199

[43] Zhang W, Gao B, Tang J, et al. Neuro-inspired computing chips. Nature Electronics. 2020;**3**:371-382. DOI: 10.1038/s41928-020-0435-7

[44] Zidan MA, Fahmy HAH, Hussain MM, Salama KN. Memristor-based memory: The sneak paths problem and solutions. Microelectronics Journal. 2013;**44**(2):176-183

[45] Kim W et al. 3-bit multilevel switching by deep reset phenomenon in Pt/W/TaOx/Pt-ReRAM devices. IEEE Electron Device Letters. 2016;**37**(5): 564-567. DOI: 10.1109/LED.2016.2542879

[46] Aziza H, Postel-Pellerin J, Moreau M. STATE: A test structure for rapid and reliable prediction of resistive RAM endurance. IEEE Transactions on Device and Materials Reliability. 2022; **22**(4):500-505. DOI: 10.1109/TDMR.2022.3213191

[47] Balatti S et al. Voltage-controlled cycling endurance of HfOx-based resistive-switching memory. IEEE Transactions on Electron Devices. 2015; **62**(10):3365-3372

[48] Aziza H, Hamdioui S, Fieback M, Taouil M, Moreau M, Girard P, et al. Multi-level control of resistive ram (rram) using a write termination to achieve 4 bits/cell in high resistance state. Electronics. 2021;**10**(18):2222

[49] Aziza H, Coulié K, Rahajandraibe W. Design considerations towards zero-variability resistive RAMs in HRS state. In: 2021 IEEE 22nd Latin American Test Symposium (LATS). IEEE; 2021

[50] Li H et al. Write disturb analyses on half-selected cells of cross-point RRAM arrays. In: 2014 IEEE International Reliability Physics Symposium. Waikoloa, HI, USA: IEEE; 2014. pp. MY.3.1-MY.3.4. DOI: 10.1109/IRPS.2014.686115

[51] Lanza M et al. Standards for the characterization of endurance in resistive switching devices. ACS Nano. 2021;**15**(11):17214-17231

[52] Rana A, Akbar T, Ismail M, et al. Endurance and cycle-to-cycle uniformity improvement in tri-layered CeO2/Ti/CeO2 resistive switching devices by changing top electrode material. Scientific Reports. 2017;7:39539

[53] Fieback M, Medeiros GC, Wu L, Aziza H, Bishnoi R, Taouil M, et al. Defects, fault modeling, and test development framework for RRAMs. ACM Journal on Emerging Technologies in Computing Systems (JETC). 2022; **18**(3):1-26

[54] Fieback M et al. Device-aware test: A new test approach towards DPPB level. In: 2019 IEEE International Test Conference (ITC). Washington, DC, USA: IEEE; 2019. pp. 1-10. DOI: 10.1109/ITC44170.2019.9000134

[55] Fieback M, Medeiros GC, Gebregiorgis A, Aziza H, Taouil M, Hamdioui S. Intermittent undefined state fault in RRAMs. In: 2021 IEEE European Test Symposium (ETS). IEEE; 2021. pp. 1-6

[56] Castellani-Coulie K, Toure G, Portal JM, Ginez O, Aziza H, Lesea A. Circuit effect on collection mechanisms involved in single event phenomena: Application to the response of a NMOS transistor in a 90 nm SRAM cell. IEEE Transactions on Nuclear Science. 2011; **58**(3):870-876

[57] Castellani-Coulié K, Aziza H, Rahajandraibe W, Micolau G, Portal JM. Development of a CMOS oscillator concept for particle detection and tracking. IEEE Transactions on Nuclear Science. 2013;**60**(4):2450-2455. DOI: 10.1109/TNS.2013.2254723

[58] Coulié K, Aziza H, Rahajandraibe W. Investigation of single event effects in a resistive RAM memory array by coupling TCAD and SPICE simulations. Journal of Electronic Testing (Springer). 2023:1-14. DOI: 10.1007/s10836-023-06068-5

[59] Castellani-Coulié K, Bocquet M, Aziza H, Portal JM, Rahajandraibe W, Muller C. SPICE level analysis of single event effects in an OxRRAM cell. In: 2013 14th Latin American Test Workshop-LATW. IEEE; 2013. pp. 1-5

Chapter 4

AI-Driven Memristor-Based Microchip Design: A Comprehensive Study

Deepthi Anirudhan Jayadevi

Abstract

Memristors represent a transformative technology with vast potential, and their integration into microchip design, aided by artificial intelligence (AI), holds the promise of revolutionizing various industries and applications. This chapter proposes the conceptual framework for the integration of AI in microchip design using memristors. It comprehensively discusses various microchip design aspects with AI, including architectural considerations, circuit design techniques, and optimization strategies employing machine learning. The chapter also delves into its potential applications in machine learning, Internet-of-Things (IoT), robotics, healthcare, etc. Ultimately, this study contributes to the development of next-generation microchips, harnessing AI and memristor technology to revolutionize computing and technological innovation.

Keywords: memristor-based integrated circuits, algorithm-level simulators, performance prediction, parameterized design, neuromorphic computing structure, artificial intelligence (AI), microchip design

1. Introduction

In the ever-evolving landscape of technology, the integration of artificial intelligence (AI) has emerged as a transformative force, propelling innovation across various domains. Among its many applications, AI's fusion with microchip design stands as a remarkable achievement, ushering in a new era of computing capabilities. AI, with its ability to process vast datasets, recognize complex patterns, and make intelligent decisions, has revolutionized the design and optimization of microchips. This transformative synergy offers profound enhancements in microchip performance, adaptability, and efficiency, redefining the boundaries of what is achievable in semiconductor technology. Central to this study is the remarkable memristor, a unique electronic component that holds the potential to reshape the microchip landscape. Memristors [1–4] exhibit characteristics that make them ideal for use in microchips, such as non-volatile memory storage and the ability to mimic synaptic behavior. These qualities enable memristors to serve as the cornerstone of advanced microchip design, facilitating innovations in memory, processing, and neuromorphic computing.

In the ceaseless pursuit of swifter, more efficient, and more intelligent computing solutions, the convergence of AI and semiconductor technology has arisen as a transformative frontier. AI-driven chip design signifies a pivotal shift in the development of microchips, where intelligent algorithms take center stage in the pursuit of optimized performance, enhanced power efficiency, and expanded functionality. In this captivating realm of innovation, memristors, a class of non-volatile resistive devices endowed with memory-like properties, have risen as indispensable components, fundamentally reshaping the microchip design landscape. The significance of AI-driven chip design cannot be emphasized enough. Traditional chip design methodologies, while reliable, are progressively falling short of meeting the escalating requirements for computational power, especially in domains such as machine learning, autonomous systems, and data-intensive computing. In this context, AI introduces an unparalleled capability to navigate expansive design spaces, unearth innovative architectures, and optimize microchip parameters to levels previously deemed unreachable. At the heart of this transformative journey lies the memristor, a device distinguished by its capacity to store and process information, erasing the distinctions between memory and logic. Memristors have reignited the quest for neuromorphic computing, providing a pathway to replicate the remarkable computational prowess of the human brain. This groundbreaking technology pledges not only to enhance the efficiency and speed of conventional microchips but also unlocks the potential for entirely novel computing paradigms.

Let us delve into the problem statement tackled in this chapter, framed as a comprehensive study. While AI-driven memristor-based microchip design holds immense promise, it also presents a myriad of intricate challenges and inquiries that require rigorous examination and analysis. These encompass inquiries surrounding the seamless integration of memristors into established chip architectures, the creation of efficient AI-driven design methodologies, and the thorough validation of performance enhancements. Addressing these complexities necessitates a thorough exploration and comprehension of the issue at hand.

Let us also explore the research objectives covered in this chapter. The principal aim of this chapter is to conduct a comprehensive study of AI-driven memristor-based microchip design, shedding light on the mutually beneficial relationship between AI algorithms and memristor technology. The primary research objectives encompass exploring memristor fundamentals and analyzing AI integration in microchip design. The memristor fundamentals are intended to provide an in-depth comprehension of memristor devices, elucidating their operational principles, and showcasing their potential applications in microchip design. Analyzing AI integration in microchip design involves evaluating the role of AI in optimizing microchip architectures, delving into neural architecture search (NAS), and dissecting other AI-driven methodologies.

This chapter embarks on a comprehensive exploration of AI-driven microchip design, specifically focusing on the integration of memristor technology. In the upcoming pages, we will delve into the fundamentals of memristor technology, exploring its characteristics, behavior, and potential applications. Subsequently, we will investigate the pivotal role of AI in optimizing microchip design and how it synergizes with memristors to unlock new possibilities in computing and technological innovation. The fusion of AI and memristor technologies, unveiling their transformative potential within the realm of microchip design are discussed. The chapter also includes a concise overview of AI-driven chip design and underscores the significance of memristors. Following that, the problem statement is presented along with the outline of the research objectives.

2. Advancements in microchip design

Microchip design has experienced remarkable advancements in recent years, driven by relentless innovation and the pursuit of higher performance, energy efficiency, and miniaturization. In this section, we delve into the current landscape of microchip design, exploring recent developments, innovations, and trends that have shaped the semiconductor industry [5]. Let's delve briefly into some of the key developments based on the relevant factors.

2.1 Processing power and performance

Quantum computing: Research and development in quantum computing have made significant strides, with quantum chips capable of solving complex problems exponentially faster than classical computers. This holds immense promise for cryptography, materials science, and optimization.

Advanced process nodes: Semiconductor foundries have introduced increasingly smaller process nodes, such as 7 nm and 5 nm, enabling the creation of chips with higher transistor densities and lower power consumption. This has revolutionized computing capabilities in various domains.

2.2 Energy efficiency

Heterogeneous integration: The integration of diverse materials and components on a single chip has led to energy-efficient designs. For example, 3D chip stacking combines various technologies (e.g., central processing units (CPUs), GPUs, memory) on a single package, reducing energy consumption and improving performance.

Low-power design techniques: Innovative low-power design methodologies, including power gating, dynamic voltage and frequency scaling (DVFS), and near-threshold voltage computing, have been employed to maximize energy efficiency in microchips.

2.3 Miniaturization and integration

System-on-chip (SoC): Advances in SoC design have resulted in highly integrated microchips that combine multiple functions on a single chip. This has paved the way for smaller and more power-efficient devices, from smartphones to IoT sensors.

Memristor integration: Incorporating memristors in microchips for in-memory computing and non-volatile memory has enhanced miniaturization and improved data processing speed and energy efficiency.

2.4 Real-world applications

Consumer electronics: Smartphones with AI-driven processors, high-resolution cameras, and energy-efficient components have become indispensable tools for communication and entertainment.

Scientific research: High-performance microchips power supercomputers used in scientific simulations, enabling breakthroughs in climate modeling, drug discovery, and astrophysics.

Healthcare: Microchips are crucial in medical imaging devices, genetic sequencing machines, and implantable medical devices, enhancing diagnostics, treatment, and patient care.

Autonomous vehicles: Advanced microchips are at the core of self-driving cars, enabling real-time perception, decision-making, and control, which are essential for safe and efficient autonomous navigation.

These advancements in microchip design have revolutionized not only industries but also opened up new frontiers for innovation. They serve as a foundation for the integration of AI and memristor technologies, offering even greater potential for future microchip development.

3. Memristors – Basics, design, and applications

Memristors, a portmanteau of "memory" and "resistor," represent a class of emerging electronic devices with unique characteristics. In this section, we delve into the fundamental principles, design considerations, and diverse applications of memristors [6].

Memristors are two-terminal non-volatile memory devices that store and process data. At their core, memristors consist of a thin insulating layer sandwiched between two metal electrodes. The key element is a memristive material, often a transition metal oxide, within the insulator. As in **Figure 1** [7], the memristor undergoes a forming process wherein the soft breakdown in the insulating layer causes it to switch from IHRS (initial high-resistance state) to LRS (low-resistance state). Subsequently, it enters a reset process from LRS to HRS (high-resistance state) and vice-versa (HRS to LRS) resulting in the set process based on the applied voltage across the metal electrodes.

The resistive switching behavior [8] due to the migration of oxygen ions was observed in Pt/ZnO/Pt structure in real-time under transmission electron microscopy (TEM) as in **Figure 2** [9].

Memristors exhibit memristance, a property wherein their resistance changes based on the history of the applied voltage. When a voltage is applied across the electrodes, oxygen vacancies in the memristive material migrate, altering the resistance.

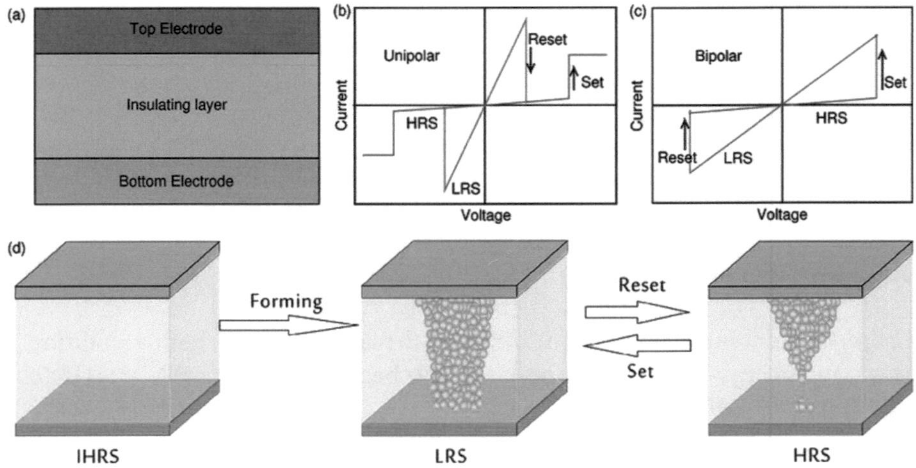

Figure 1.
Illustration - structure of a Memristor [7] (a) schematic illustration of metal-insulator-metal (MIM) structure for resistive switching memristor, (b) I-V characteristics for memristors operated in unipolar (or nonpolar) mode and (c) in bipolar mode, and (d) schematic illustration of the resistive switching process in the filament-type memristors.

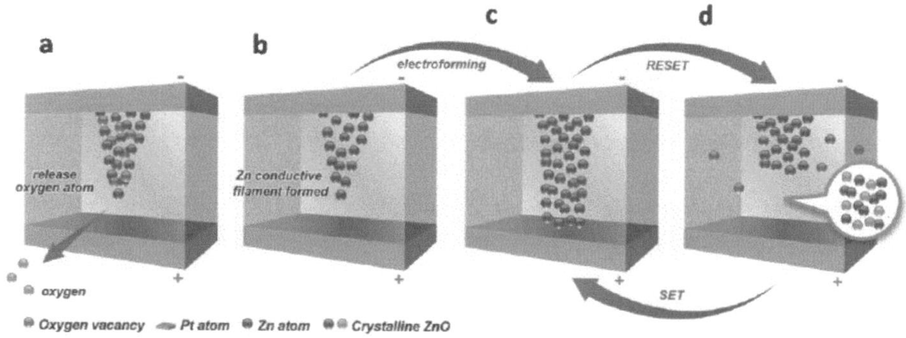

Figure 2.
Schematic depicting the operation of a memristor showing the migration of oxygen vacancies (the mechanism of resistive switching in Pt/ZnO/Pt devices) [9] (a) the release of oxygen gas (O_2) leads to the oxygen vacancies in the bulk of ZnO. (b) the migration of the mobile oxygen vacancies toward the cathode (oxygen ions (O_{2-}) toward the anode) and the rearrangement of Zn-dominated ZnO_{1-x}. (c) the precipitation of Zn atoms forms a conductive filament. (d) the rupture of the filament. When the energy of joule heating is provided to the thermochemical reaction, the filament will rupture and change back to ZnO. Owing to the migration of oxygen ions, the ReRAM resets back to the off state. These memristive devices can be classified as filament-type, interface-type, and bulk-type based on their active switching region and their geometry [10].

This migration depends on the polarity and magnitude of the applied voltage. Memristors possess several unique attributes like non-volatility, high speed, scalability, and analog and digital operation. Non-volatility implies that memristors retain their resistance state even when power is removed. Memristors are characterized by high speed which enables them to switch rapidly, making them suitable for memory and logic applications. Scalability is another remarkable feature that can be miniaturized to nanoscale dimensions, facilitating integration into microchips. Memristors ensure both analog and digital operation which enables them to function as both memory storage elements and synaptic connections in neuromorphic systems.

Various materials exhibit memristive behavior, including transition metal oxides like titanium dioxide (TiO_2), hafnium oxide (HfO_2), and tantalum oxide (TaOx). The choice of memristive material influences device characteristics. Memristor devices can have diverse architectures, such as crossbar arrays, where memristors intersect at crosspoints. These arrays are integral to neuromorphic computing and in-memory computing applications. Some applications where memristors show promising results are as follows:

Memory storage: Memristors are employed in resistive random-access memory (RRAM) for non-volatile storage. Their speed, scalability, and endurance make them promising candidates for future memory technologies.

Neuromorphic computing: Memristors are crucial in neuromorphic systems, where they emulate synapses. This enables energy-efficient, brain-inspired computing for tasks like pattern recognition and machine learning.

In-Memory computing: Memristors are used for in-memory computing, where computation occurs within memory, reducing data transfer bottlenecks and improving processing speed.

Energy-efficient AI acceleration: They enhance AI model training and inference speed while reducing power consumption, making them valuable in AI accelerators.

IoT and edge devices: In energy-constrained IoT devices, memristors offer low-power storage and computation capabilities.

Reconfigurable electronics: Memristors enable reconfigurable electronics, adapting circuitry for specific tasks.

Quantum Computing: Some memristive devices show promise in quantum computing applications.

4. Memristor-based microchip design

Several research groups and companies have developed microchip prototypes integrating memristors. These prototypes often focus on enhancing memory and computing capabilities. In this section, we delve into the current state of memristor-based microchip design, showcasing prototypes, products, and notable advancements to highlight the progress in this exciting field.

Performance metrics	Traditional microchips	Memristor-based microchips
Processing speed	Traditional microchips, such as those based on CMOS technology, offer high processing speeds, especially for tasks involving sequential processing.	Memristor-based microchips can potentially offer similar processing speeds for many tasks. However, their advantage lies in certain parallel and memory-intensive tasks due to in-memory computing capabilities, which can lead to significantly faster processing times.
Power efficiency	Traditional microchips can be power-hungry, especially when operating at high clock speeds. Power efficiency is a concern, particularly in battery-powered devices.	Memristor-based microchips have the potential for superior power efficiency. They can perform certain computations with lower energy consumption, making them suitable for energy-constrained applications like IoT devices.
Area utilization	Traditional microchips have fixed architectures, and their area utilization depends on the complexity of the design. Smaller transistors in advanced nodes allow for more components in the same space.	Memristor-based microchips can optimize area utilization by adapting their architecture to specific tasks. This adaptability can lead to smaller, more compact microchips without sacrificing performance.
Memory and storage	Traditional microchips often rely on separate memory and storage components (e.g., RAM, NAND flash). This can lead to data transfer bottlenecks.	Memristor-based microchips can integrate memory and storage, providing a significant advantage for in-memory computing. This reduces data transfer latency and improves overall system performance.
Scalability	Traditional microchips have been subject to Moore's law, which predicts a doubling of transistor counts approximately every two years. This has driven advancements in performance.	Memristor-based microchips also benefit from scaling, but their advantage lies in adapting their architecture to different scales and applications, potentially reducing the need for extensive redesign.
Specialized architectures	Traditional microchips are designed with general-purpose architectures, making them suitable for a wide range of tasks.	Memristor-based microchips can be tailored to specific tasks and may exhibit superior performance in specialized applications, such as neuromorphic computing.

Table 1.
Performance metrics comparison of memristor-based microchips with traditional microchips.

Let us draw a comparison of the performance metrics of memristor-based microchips with traditional microchips involving assessing various factors. It is important to note that the specific metrics can vary depending on the application and use case. Here, we will compare them in terms of several key performance metrics based on the inferences [11] made from experimental results of different research works (**Table 1**).

It's important to note that the comparison between memristor-based and traditional microchips can vary depending on the specific implementation, technology node, and design choices. Memristor-based microchips hold promise for certain applications where their unique characteristics, such as in-memory computing and adaptability, provide a competitive advantage in terms of performance and efficiency. **Figure 3** shows the 2 cm×2cm silicon microchips that have been designed by means of Synopsys software and fabricated in a 200 mm silicon wafer in an industrial clean room using a 180nm CMOS technology node [12].

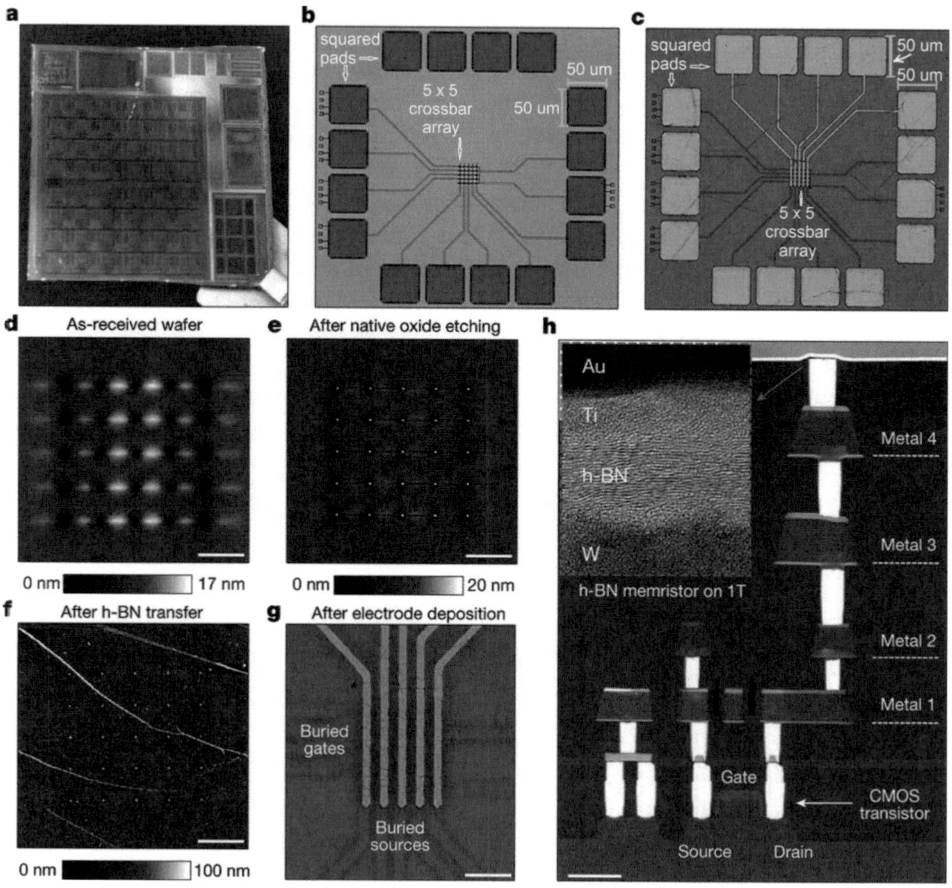

Figure 3.
Fabrication of hybrid 2D–CMOS memristive microchips. (a) Photograph of the 2 cm × 2cm microchips containing the CMOS circuitry. b,c, optical microscope images of a part of the microchip containing a 5 × 5 crossbar array of 1 T1M cells, as received (b) and after fabrication (c). The size of the squared pads is 50 μm × 50 μm. d–f, topographic maps collected with atomic force microscopy of the vias in the 5 × 5 crossbar arrays on the wafers as received (d), after native-oxide etching (e) and after the transfer of the h-BN sheet (f). (g) optical microscope image of a finished 5 × 5 crossbar array of 1T1M, that is, after h-BN transfer and top electrodes deposition. (h) high-angle annular dark-field cross-sectional scanning transmission electron microscope image of a 1T1M cell in the crossbar array. The inset, which is 20 nm × 16nm, shows a cross-sectional TEM image of the Au–Ti–h-BN–W memristor on the via; the correct layered structure of h-BN can be seen. Scale bars, d–f, 10 μm; g, 25 μm; h, 600 nm. [12].

5. AI in memristor-based microchips – Proposed future perspectives

Memristors have proven themselves to combine their ability for memory and computation in realizing artificial synapses essential to building neuromorphic hardware and thereby influencing the AI era. The integration of memristors in AI can lead to significant advancements in machine learning with better implementations since the memristor electronics are more accurate and energy-efficient than existing AI hardware. They can pack huge amounts of computing power into hand-held devices, removing the need to be connected to the internet and allowing data processing to be local rather than sending it to remote data centers. In addition to neuromorphic computing, AI integration has revolutionized memristor technology, enabling advanced applications in in-memory computing, energy-efficient IoT devices, and real-time adaptive systems [13, 14].

On the contrary, AI has proven itself to play a pivotal role in microchip design, even halving the cost of designing steps in microchip fabrication [15]. AI is capable of assisting in reducing design time, improving efficiency, and developing more robust and reliable designs. A future direction for memristive microchip design might include AI-enhanced design techniques for realizing memristor-based microchips. AI-driven memristor microchip design involves the use of AI technologies such as machine learning in the tool flow to design, verify, and test semiconductor devices. One typical proposal for the framework at a conceptual level is the implementation of reinforcement algorithms that can pave the way to highly optimized memristive design and implementation as in **Figure 4** wherein both the design optimization and fabrication constraints are supposed to be taken into consideration. It has been observed that architectural modifications can contribute to circuit reliability in memristor crossbar arrays (MCAs) [16]. Incorporating AI-driven techniques into memristor-based microchip design enhances not only efficiency and cost-effectiveness but also plays a significant role in improving circuit reliability. Here, we will delve into the role of AI in enhancing circuit reliability in the context of memristor-based microchip design.

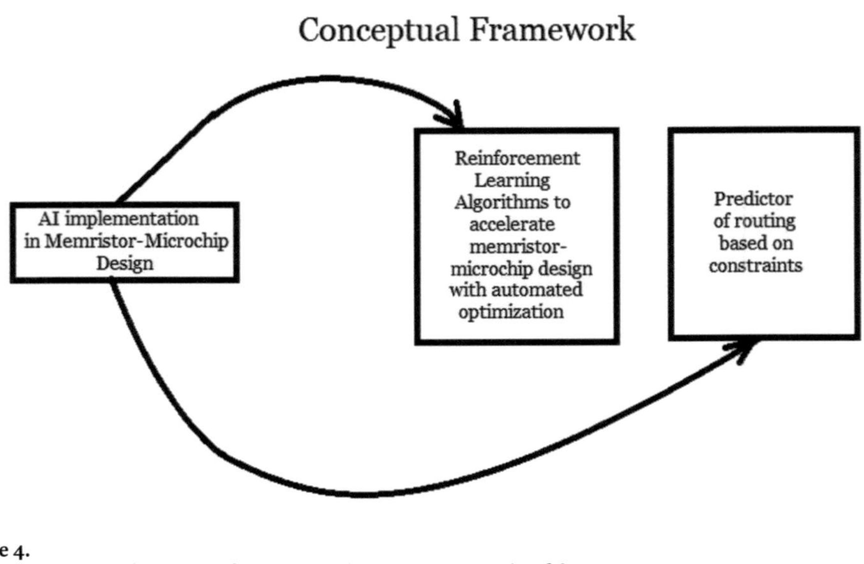

Figure 4.
Proposed conceptual framework for AI-assisted memristor microchip fabrication.

5.1 AI-driven circuit reliability enhancement in memristor-based microchips

AI-driven memristor-based microchip design involves the use of AI technologies, including machine learning and reinforcement algorithms, to optimize various aspects of design, verification, and testing. One of the critical areas where AI can have a profound impact is in improving circuit reliability. Here are some key aspects:

Fault tolerance and self-healing: AI algorithms can continuously monitor the operation of memristor-based microchips and identify potential faults or errors. In case of detected faults, AI-driven systems can employ self-healing mechanisms by rerouting signals, bypassing defective components, or implementing redundancy strategies to ensure uninterrupted operation.

Predictive maintenance: AI can analyze operational data from memristor-based microchips and predict potential failures or degradation. This enables proactive maintenance and replacement of components, reducing the risk of sudden chip failures and improving overall reliability.

Adaptive redundancy management: AI algorithms can dynamically manage redundancy within the microchip. In the event of component failures, AI can reroute signals through redundant pathways, ensuring that the chip continues to function reliably.

Heat and power optimization: AI-driven microchip design can optimize power consumption and heat dissipation. By intelligently managing power usage and heat generation, AI can prevent overheating and ensure the chip's reliability over its lifespan.

Error correction and resilience: AI-enhanced error correction codes and algorithms can identify and rectify data errors. This improves the overall data integrity and reliability of memristor-based microchips.

Real-time monitoring: AI systems can provide real-time monitoring of the microchip's operating conditions. Any deviations or anomalies can be detected and addressed promptly, minimizing the risk of catastrophic failures.

5.2 Microchip design – Adaptable recent trends in memristor-based microchip design

The landscape of microchip design is in a constant state of evolution, driven by the unceasing demand for increased computational power, energy efficiency, and versatile functionality. In recent years, the convergence of microchip design and AI has emerged as a formidable force, reshaping the industry and opening new horizons in chip architecture and performance. This section provides an overview of the latest trends in microchip design and highlights the transformative impact of AI-driven optimizations. Let us explore the recent trends in microchip design.

Miniaturization and Moore's law: Microchip manufacturers continue to push the boundaries of miniaturization, adhering to Moore's law, which predicts a doubling of transistor density approximately every two years. This relentless scaling has enabled the creation of smaller, more powerful chips.

Specialized hardware acceleration: To meet the demands of AI and machine learning applications, microchip designers are increasingly incorporating specialized hardware accelerators such as GPUs (graphics processing units) and TPUs (tensor processing units). These accelerators enhance the performance of AI workloads significantly.

Heterogeneous integration: Modern microchips often incorporate a diverse range of components, including CPUs, GPUs, memory, and AI accelerators. Heterogeneous integration allows for optimized performance in a wide array of applications.

Energy efficiency: With concerns over power consumption and environmental impact, microchip designers are placing a strong emphasis on energy efficiency. This involves optimizing not only the hardware but also developing advanced power management techniques.

Security: The rise in cyber threats has necessitated the integration of robust security features directly into microchips. This includes hardware-based encryption and secure boot processes.

The integration of AI techniques in microchip design has opened up exciting possibilities and yielded tangible benefits. Here are a few examples:

AI-driven layout optimization: AI algorithms can optimize the physical layout of components on a microchip, reducing signal interference and improving performance.

Neural architecture search (NAS): NAS techniques use AI to automatically search for the best microchip architecture for a given task. This approach has led to the creation of highly efficient, specialized chips for AI workloads.

Predictive maintenance: AI can be used to predict when a microchip is likely to fail, enabling proactive maintenance and reducing downtime.

AI-based testing: Machine learning algorithms can enhance the testing process by identifying defects more accurately and quickly, improving chip yield.

Autonomous chip design: In some cases, AI is being used to autonomously design microchips. This has the potential to accelerate the design process and lead to more innovative solutions.

The fusion of AI and microchip design represents a powerful synergy that is reshaping industries ranging from healthcare and autonomous vehicles to data centers and consumer electronics. As we delve deeper into the integration of memristor technology with AI-driven design in the subsequent sections, it becomes clear that these advancements hold the promise of unlocking entirely new possibilities in microchip functionality and efficiency.

Memristors, characterized by their unique attributes, have emerged as a disruptive force in microchip design, offering the potential for significant performance and functionality enhancements. This section provides an in-depth exploration of how memristors can contribute to microchip design, discussing both their potential advantages and the challenges associated with their integration. Enhancing microchip performance with memristors is of significant interest. Let us delve into some of the measures taken.

Non-volatile memory: Memristors excel as non-volatile memory elements, retaining data even when power is removed. This attribute is invaluable in applications where data integrity is paramount, such as servers, embedded systems, and IoT devices.

In-memory computing: Leveraging memristors for in-memory computing, where data storage and processing coexist, reduces the need for data transfers between memory and processing units. The result is swifter computation and reduced power consumption, particularly advantageous in AI applications.

High-speed data storage and access: Memristors boast rapid write and read speeds, rendering them suitable for applications demanding high-speed data access, such as real-time analytics, video processing, and AI workloads.

Neuromorphic computing: With properties akin to human brain synapses, memristors are ideal candidates for neuromorphic computing. They enable the creation of artificial neural networks capable of learning and adapting, potentially revolutionizing AI applications.

Energy efficiency: Memristors typically consume minimal power during state transitions, contributing to overall energy-efficient microchip design.

The potential advantages of memristors can be summarized as follows:

Improved performance: Memristors can empower microchips to process data faster and more efficiently, a pivotal attribute in AI applications, where real-time processing of extensive data is the norm.

Reduced energy consumption: The low power requirements of memristors can result in substantial energy savings, particularly in battery-powered devices.

Enhanced memory solutions: Memristor-based memory devices like Resistive RAM (RRAM) offers compelling alternatives to traditional memory technologies due to their non-volatile nature and high-speed operation.

Neuromorphic capabilities: Memristors, mimicking synaptic behavior, open possibilities for the development of brain-inspired computing systems with learning and adaptation capabilities.

We should take note of the challenges in memristor integration as we progress.

Manufacturing consistency: Ensuring consistent manufacturing of memristor devices with reproducible properties can be challenging, impacting their reliability and performance.

Endurance and retention: Memristor-based memory technologies may encounter challenges related to endurance (how many write cycles they can endure) and data retention (how long they can retain stored data without degradation).

Integration complexity: Integrating memristors into existing microchip architectures can be complex, potentially necessitating significant design and fabrication process changes.

Compatibility: Achieving compatibility with current semiconductor manufacturing processes can be a hurdle, possibly requiring specialized fabrication techniques.

Despite these challenges, the promise of memristor technology in microchip design is unmistakable. The research and development efforts are addressing many of these challenges, paving the way for a new era of microchip innovation [17–19]. In the ensuing sections, we dive deeper into the synergy between AI-driven methodologies and memristor-based microchip design, exploring the potential for groundbreaking advancements.

The amalgamation of AI algorithms with memristor-based microchip design forms a potent alliance poised to reshape the semiconductor industry [20]. In this section, we explore how AI algorithms can elevate the design process and provide an overview of pivotal AI-driven optimization methods, including neural architecture search (NAS). It is fascinating to explore how AI algorithms enhance the design process [21].

Exploring design spaces: AI algorithms possess the capacity to traverse vast design spaces, a task impractical or impossible for human designers to undertake manually. This becomes particularly valuable when optimizing the intricate architecture of memristor-based microchips.

Optimizing parameters: AI excels in fine-tuning a myriad of parameters, encompassing the arrangement of memristor devices, interconnects, voltage levels, and more, all to maximize chip performance. This results in highly optimized designs that may remain elusive through conventional methods.

Reducing development time: AI-driven design significantly expedites the development cycle by automating numerous aspects of the design process. This acceleration translates into quicker time-to-market for innovative microchip creations.

Adaptive learning: AI algorithms demonstrate the ability to adapt and learn from previous design iterations, continuously enhancing their effectiveness. They can incorporate feedback and real-world chip implementation data, which in turn informs and improves future designs.

Let us delve into the overview of neural architecture search (NAS) and AI-driven optimization. NAS stands out as a prominent AI-driven optimization method that has garnered substantial attention in the microchip design domain. NAS leverages neural networks to automatically uncover optimal microchip architectures. Here is an overview of its workings:

Search space definition: NAS initiates by defining a search space encompassing various possible chip architectures. This space spans diverse design choices, including the configuration of memristor devices, interconnects, and other architectural elements.

Neural network controller: A neural network-based controller undergoes training to generate and evaluate different chip architectures within the defined search space. This controller learns to predict the performance of each architecture based on specified objectives, such as speed, power efficiency, and area.

Architecture sampling: The controller proceeds to sample various chip architectures based on its performance predictions. These sampled architectures undergo evaluation, typically through simulation or prototyping.

Feedback loop: Performance data from the evaluated architectures is fed back into the neural network controller. This feedback loop enables the controller to learn from the results, refining its predictions and architectural choices over successive iterations.

Optimal architecture selection: Eventually, the NAS process converges to identify the architecture that best aligns with the specified objectives. This chosen architecture undergoes further refinement and implementation.

Beyond NAS, various other AI-driven optimization methods, such as genetic algorithms, reinforcement learning, and Bayesian optimization, have found application in memristor-based microchip design. These methods exhibit adaptability to diverse design challenges and objectives, rendering them versatile tools in the quest for optimized chip performance.

In the subsequent sections, we delve deeper into the empirical results of AI-driven memristor-based microchip design, exemplifying the tangible benefits and innovations that this synergy brings to the semiconductor technology landscape.

5.3 Experimental framework – A study

In the pursuit of exploring the synergistic potential of AI-driven memristor-based microchip design, it is imperative to establish a robust experimental framework. This section provides an in-depth account of the experimental environment and the methodology employed for training and assessing AI models in the context of microchip design. The commonly used experimental environment and tools are explored below.

Hardware platform: The experimental setup typically comprises a high-performance computing cluster or cloud-based infrastructure equipped with potent CPUs and GPUs. These resources are indispensable for efficiently training and evaluating AI models.

Software tools: A variety of software tools and libraries are enlisted for AI-driven microchip design, including TensorFlow, PyTorch, or other deep learning frameworks for constructing and training neural networks. Additionally, specialized simulation software may be employed for accurately modeling the behavior of memristors in chip designs.

Memristor prototyping: In cases where physical memristor-based chips are involved, specialized hardware and fabrication facilities are indispensable. This encompasses access to cleanroom facilities for chip fabrication and tools for testing and characterizing memristor devices.

Datasets: Datasets containing pertinent information for microchip design are utilized. These datasets may encompass memristor characteristics, performance metrics, and historical data from previous chip designs.

Let us now understand the training and evaluation of AI models.

Data preparation: The initial step encompasses collecting and preprocessing data. This includes the preparation of datasets for training and validation, ensuring their fidelity in representing the problem space and the characteristics of memristor devices.

Model development: AI models, often neural networks, are meticulously crafted to optimize specific aspects of microchip design. These models may encompass convolutional neural networks (CNNs) for tasks such as image recognition (e.g., identifying patterns in memristor behavior) or recurrent neural networks (RNNs) for tasks involving time-series analysis.

Training process: The AI models are subjected to rigorous training using the prepared datasets. During this training phase, the models assimilate the ability to make predictions or decisions based on input data, progressively refining their performance through iterations. This process may necessitate substantial computational resources and exhibit varying durations contingent on the complexity of the task.

Validation and testing: The trained models undergo evaluation on separate validation datasets to gauge their performance. Various metrics pertinent to microchip design objectives, encompassing speed, power efficiency, and area utilization, are enlisted to quantify performance.

Hyperparameter tuning: To optimize model performance, hyperparameter tuning techniques come into play. This entails the adjustment of parameters such as learning rates, batch sizes, and network architectures to fine-tune the model's behavior.

Cross-validation: Cross-validation techniques may be harnessed to ensure the model's generalizability and robustness. This technique involves partitioning the data into multiple subsets for both training and validation, facilitating the detection of overfitting or underfitting issues.

Real-world testing: In some instances, the trained AI models find practical implementation in real-world memristor-based microchip designs. Their performance is then meticulously evaluated in practical applications to validate their efficacy.

The experimental setup and methodology play a pivotal role in empirically validating AI-driven memristor-based microchip design. Through rigorous experimentation and validation, the capabilities and potential advantages of AI-driven design methodologies, in conjunction with memristor technology, can be quantified and demonstrated. The enhanced performance, architectural innovations, and potential for neuromorphic computing can open doors to groundbreaking advancements. Nevertheless, addressing limitations and potential sources of error remains an ongoing endeavor as this field continues to evolve. The fusion of AI and memristor technology represents a potent force that will shape the future of microchip design and redefine the capabilities of electronic systems. The exciting possibilities and innovations that AI-driven memristor-based microchip design may bring to various domains, from AI itself to IoT, energy harvesting, security, ethics, and more are yet to be explored. Let us delve into the potential future directions for specific application areas in AI-driven memristor-based microchip design.

Advanced neuromorphic systems: The development of more advanced neuromorphic systems is a promising direction. Research in this area may lead to memristor-based microchips that closely mimic the intricate functions of the human brain, opening new frontiers in AI and cognitive computing.

Customized AI accelerators: Tailoring AI accelerators within microchips to specific tasks and applications is an avenue with immense potential. These customized accelerators can lead to higher efficiency and better performance for specialized tasks like natural language processing or image recognition.

Energy harvesting: Exploring energy harvesting techniques that leverage memristor-based microchips could lead to self-powering IoT devices and sensors. This research could revolutionize energy-efficient electronics.

Security and privacy: Investigating the security and privacy implications of AI-driven memristor-based microchips is crucial. Research should focus on addressing potential vulnerabilities and developing robust encryption and security measures.

Ethical considerations: Ethical considerations regarding the use of AI-driven technology in microchips must be explored further. This includes addressing bias in AI algorithms and ensuring responsible and transparent AI-driven design practices.

Let us discuss the challenges related to manufacturing that need to be overcome for the widespread adoption of AI-driven memristor-based microchips.

Scalability: Achieving scalable manufacturing processes for memristor-based microchips is a significant challenge. Consistency and reliability must be maintained as chip sizes increase and production scales up.

Cost constraints: The cost of producing memristor-based microchips needs to be competitive with traditional chip manufacturing. Reducing production costs while maintaining quality is a critical challenge.

Standardization: Establishing industry standards for AI-driven memristor-based microchips is essential to ensure compatibility and interoperability. Standardization efforts must be coordinated to avoid fragmentation.

Regulatory compliance: Navigating regulatory frameworks and ensuring compliance with safety and environmental regulations is a complex challenge that must be addressed for market acceptance.

Education and skill development: Widespread adoption also depends on a skilled workforce capable of designing, manufacturing, and maintaining these innovative chips. Investing in education and training programs is crucial.

Long-term reliability: Ensuring the long-term reliability and durability of memristor-based microchips is essential for their adoption in critical applications such as aerospace and healthcare.

In conclusion, the future of AI-driven memristor-based microchip design holds tremendous promise, from advanced neuromorphic computing to energy-efficient IoT devices. However, addressing manufacturing scalability, cost constraints, standardization, regulatory compliance, and other challenges is imperative to facilitate widespread adoption and maximize the potential of this groundbreaking technology. Research and innovation in these areas will continue to shape the trajectory of the semiconductor industry in the years to come.

6. Conclusion

In this chapter, we embarked on a transformative journey through the realm of AI-driven memristor-based microchip design, uncovering a landscape of innovation poised to reshape the semiconductor industry. Our exploration began with a deep dive into the fundamentals of memristor technology, exploring its unique attributes and its potential applications in memory and computing. From there, we ventured into the current state-of-the-art microchip design, elucidating the integration of

AI techniques, highlighting recent trends, and showcasing the power of AI-driven optimizations in chip architecture.

The heart of this chapter was the pivotal role of memristors in microchip design. We unraveled how memristors elevate microchip performance, offering non-volatile memory solutions, enabling in-memory computing, and fostering neuromorphic capabilities. As we examined the landscape, we recognized both the potential advantages and the complexities that memristor technology introduces to the field.

The spotlight then shifted to the synergy between AI algorithms and memristor technology, a compelling combination where we meticulously outlined how AI algorithms enhance the design process, from navigating intricate design spaces to optimizing critical parameters. This dynamic duo accelerates development cycles, nurtures adaptability, and presents game-changing tools like NAS in the microchip designer's arsenal. We underscored the paramount importance of ethical considerations and robust security measures in this transformative field.

Looking ahead, we cast our gaze toward the future, spotlighting potential pathways for further research and development. Advanced neuromorphic systems beckon, promising microchips that mimic the intricacies of the human brain. Customized AI accelerators within microchips hold the potential to elevate efficiency and performance for specialized tasks. Exploring energy harvesting techniques fueled by memristor-based microchips could revolutionize the world of energy-efficient electronics. Ethical considerations and security imperatives must continue to be at the forefront as we navigate this transformative landscape.

Yet, challenges stand as gatekeepers to the widespread adoption of memristors in AI-driven microchip technology. Manufacturing scalability looms as a significant hurdle, demanding consistent reliability as chip sizes increase and production scales. Cost constraints must be met, ensuring competitive production costs while preserving quality. Standardization becomes imperative, fostering compatibility, and interoperability while averting fragmentation. Navigating complex regulatory frameworks and ensuring compliance is a formidable task. A skilled workforce, educated and equipped to design and maintain these innovative chips, remains essential. Finally, ensuring the long-term reliability and durability of memristor-based microchips is paramount for their deployment in critical applications like aerospace and healthcare.

In conclusion, this chapter has unveiled the conceptual framework that can pronounce the profound significance of AI-driven microchip design using memristors as a catalyst for innovation in the semiconductor industry. It marks the convergence of cutting-edge technologies that promise enhanced performance, unmatched adaptability, and the tantalizing potential of neuromorphic capabilities. These implications extend across diverse domains, from the smallest edge devices to the massive data centers powering our digital world. As we confront the challenges that lay ahead, they are not roadblocks but opportunities for pioneering research, collaborative endeavors, and a collective push toward unlocking new horizons in the realms of electronics and computing. The synergy between AI and memristor technology stands poised to define the future of microchip design, setting the stage for a transformative era in the world of technology and innovation.

Acknowledgements

I express my sincere thanks to the invitation from IntechOpen to contribute to this esteemed book edited by Dr. Yao-Feng Chang from The University of Texas at Austin.

Author details

Deepthi Anirudhan Jayadevi
Intel OneAPI Innovator, IICSE University, United States

*Address all correspondence to: deepthi_aj2002@yahoo.com

References

[1] Chua LO. Memristor—The missing circuit element. IEEE Transactions on Circuit Theory. 1971;**18**(5):507-519

[2] Yang JJ, Strukov DB, Stewart DR. Memristive devices for computing. Nature Nanotechnology. 2013;**8**(1):13-24

[3] Xia Q, Robinett W, Cumbie MW, Banerjee N, Cardinali TJ, Tour JM. Memristor-CMOS hybrid integrated circuits for reconfigurable logic. Nano Letters. 2009;**9**(10):3640-3645

[4] Isah A, Bilbault J-M. Review on the basic circuit elements and memristor interpretation: Analysis, technology and applications. Journal of Low Power Electronics and Applications. 2022;**12**(3):44. DOI: 10.3390/jlpea12030044

[5] Burr GW, Shelby RM, Sebastian A, Kim S, Kim S, Sidler S, et al. Recent progress in phase-change memory technology. IEEE Journal of Solid-State Circuits. 2015;**50**(1):202-219

[6] Mazumder P, Kang S-M, Waser R. Memristors: Devices, models, and applications. In: Proceedings of the IEEE. Vol. 100. No. 6. 2012. pp. 1911-1919

[7] Xiao Y, Jiang B, Zhang Z, Ke S, Jin Y, Wen X, et al. A review of memristor: Material and structure design, device performance, applications and prospects. Science and Technology of Advanced Materials. 2023;**24**:1. DOI: 10.1080/14686996.2022.2162323

[8] Sun W, Gao B, Chi M, Xia Q, Yang JJ, Qian H, et al. Understanding memristive switching via in situ characterization and device modeling. Nature Communications. 2019;**10**(1):1-13. DOI: 10.1038/s41467-019-11411-6

[9] Hu SG, Wu S, Jia W, Yu Q, Deng L, Fu YQ, et al. Review of nanostructured resistive switching memristor and its applications. Nanoscience and Nanotechnology Letters. 2014;**6**:729-757. DOI: 10.1166/nnl.2014.1888

[10] Chen JY, Hsin CL, Huang CW, Chiu CH, Huang YT, Lin SJ, et al. Dynamic evolution of conducting nanofilament in resistive switching memories. Nano Letters. 14 Aug 2013;**13**(8):3671-3677. DOI: 10.1021/nl4015638. Epub 2013 Jul 19. PMID: 23855543

[11] Keerthy Rai V, Sakthivel R. Survey based on comparison of different CMOS logic styles with hybrid CMOS memristor logic style with issues and parameters. International Journal of Pure and Applied Mathematics. 2018;**118**(18):4373-4392

[12] Zhu K, Pazos S, Aguirre F, Shen Y, Yuan Y, Zheng W, et al. Hybrid 2D–CMOS microchips for memristive applications. Nature. 2023;**618**(7963):57-62. DOI: 10.1038/s41586-023-05973-1

[13] Mina R, Jabbour C, Sakr GE. A review of machine learning techniques in analog integrated circuit design automation. Electronics. 2022;**11**(3):435. DOI: 10.3390/electronics11030435

[14] Joksas D, Freitas P, Chai Z, Ng WH, Buckwell M, Li C, et al. Committee machines—A universal method to deal with non-idealities in memristor-based neural networks. Nature Communications. 2020;**11**(1):1-10. DOI: 10.1038/s41467-020-18098-0

[15] Wang Y, Huang M. Human–AI team halves cost of designing step in microchip fabrication. Nature.

2023;**616**(7958):667-668. DOI: 10.1038/d41586-023-01353-x

[16] Gnawali KP, Paudel BR, Mozaffari SN, Tragoudas S. Reliability enhancements in memristive neural network architectures. IEEE Transactions on Nanotechnology. 2019;**18**:866-878. DOI: 10.1109/TNANO.2019.2933806

[17] Prezioso M, Merrikh-Bayat F, Hoskins BD, Adam GC, Likharev KK, Strukov DB. Training and operation of an integrated neuromorphic network based on metal-oxide memristors. Nature. 2015;**521**(7550):61-64

[18] Kim KH, Gaba S, Wheeler D, Cruz-Albrecht JM, Hussain T, Akinwande D. A functional hybrid memristor crossbar-array/CMOS system for data storage and neuromorphic applications. Nano Letters. 2011;**12**(1):389-395

[19] Indiveri G, Linares-Barranco B, Hamilton TJ, van Schaik A, Etienne-Cummings R, Delbruck T, et al. Neuromorphic silicon neuron circuits. Frontiers in Neuroscience. 2011;**5**:73

[20] Cai F, Correll J, Lee S, Lim Y, Bothra V, Zhang Z, et al. A fully integrated reprogrammable memristor–CMOS system for efficient multiply–accumulate operations. Nature Electronics. 2019;**2**:1. DOI: 10.1038/s41928-019-0270-x

[21] Kahng AB. Machine learning applications in physical design: Recent results and directions. In: Proceedings of the 2018 International Symposium on Physical Design (ISPD '18). New York, NY, USA: Association for Computing Machinery; 2018. pp. 68-73. DOI: 10.1145/3177540.3177554

Chapter 5

Thermal Reliability Issues in ReRAM Memory Arrays

Amrita Chakraborty, Mohammad S. Al-Mamun and Marius K. Orlowski

Abstract

A localized buildup of Joules heat occurs in a memory cell as a result of the switching of resistive memory (resistive random access memory, (ReRAM)) cells. In ReRAM memory crossbar architecture, heat accumulated in one cell transfers to nearby cells via shared electrode metal lines and impacts adversely the performance of the devices. The cell-to-cell heat transfer leads to reduced number of switching cycles (SWC) and, in specific circumstances, to a loss of a bit in the cell brought about by a rupturing of the nanofilament. This chapter identifies the main causes for the thermal ReRAM reliability issues, offers a detailed analysis of the intercell heat transfer, characterizes its effects on the electric behavior of cells, and describes the thermal cross-talk in terms of three key material properties: specific heat capacity, the mass given by the geometry of the conductor material, and thermal conductivity of the electrode. By depositing several distinct composite inert electrodes, multiple ReRAM arrays have been manufactured to alter deliberately the thermal conductivity, specific heat capacity, and the electrode geometry. The proposed thermal analysis, based on those electrode properties, accurately anticipates the experimentally discovered variations in the degradation of electric output characteristics of the ReRAM devices operated under the same conditions.

Keywords: ReRAM, thermal cross-talk, thermal properties of conductors, electrical and thermal reliability, specific heat capacity, switching stability of memory cells

1. Introduction

Since scaling limitations have been reached for conventional nonvolatile memory (NVM) that relies on metal-oxide-semiconductor field-effect transistors (MOSFETs) with a floating gate, it is essential to design substitute memory cells. Owing to its simple structure, strong miniaturization capacity, low power usage, high switching frequency, large ON/OFF resistance ratio, remarkable endurance, and retention features, resistive random access memory (ReRAM) is one of the leading contenders to supplant the existing NVM technology [1–3]. For neuromorphic applications, resistive memory cells are also very promising building elements [4, 5]. An authoritative review of the memristor technologies has been given recently by YF Chang [6]. The suitability of several types of nonvolatile and volatile memristive switches for neuromorphic applications has also been thoroughly reviewed [7]. The two-terminal resistive switching device, such as $Cu/TaO_x/Pt$, consists of two metal electrodes and a solid electrolyte insulating layer (in this case, TaO_x). In response to an appropriate applied electric field, the solid electrolyte film exhibits a transition between low and high states of resistance. In ReRAM devices called bipolar resistive devices, the switchover between a high resistance state (HRS), denoted by the resistance R_{off}, and a low resistance state (LRS), denoted by the resistance R_{on}, takes place at a polarity opposite to that at which the transition between HRS and LRS occurs. The establishment and dissolution of conductive filaments (CFs) in the matrix of the insulating dielectric as a result of metal ions and charged defect electromigration as well as heat impacts account for the resistance switching effect [8]. Metallic cations of the so-called active electrode (in our case, Cu) migrate in the electric field toward the second electrode, known as the inert electrode (such as Pt, Rh, Ru, and Co), where they are reduced to neutral atoms [9]. A metal nanofilament is created between the two electrodes as a result of the Cu metal atoms gradually accumulating at the interface of Pt and TaO_x. This mechanism is known as the "SET operation." The filament ruptures at a definite current I_{res} when an opposite bias is applied to the copper electrode. A relatively large current of a few milliamperes (mA) is flowing through the filament at that point, dissipating Joules heat, and thus triggering the copper atoms' outdiffusion from the body of the nanofilament. The RESET operation is the term used to describe the filament rupturing. Joule heating is essential during the RESET processes. The dominance of Joule heating in the RESET process has been thoroughly proven in several investigations [10–15]. The R_{off}/R_{on} ratio degrades due to the self-heating effects of switching the cell, decreasing as the ambient temperature increases [16]. The computational reliability of neuromorphic computing systems based on ReRAM cells suffers noticeably from such variations in R_{on} and R_{off} of ReRAM cells [5]. In particular, a cell at higher temperature is more likely to produce an inaccurate output. ReRAM's heat stability is a separate issue for embedded memories used in automotive applications [17].

The temperature of the filament during the RESET procedure will rise, rendering the cell more unstable under the same operating conditions if the bottom electrode is thinner because less Joules heat generated in the nanofilament can be dissipated. Reference [18] provides a thorough overview of resistive switching processes. Repeated switching of a cell causes the device to accumulate more Joules heat. It has been discovered [19, 20] that the Joules heat generated in a particular cell is preferably transferred along the copper and platinum electrodes, impacting the next cell positioned along the same electrode lines and impairing its electrical characteristics. Furthermore, it has been demonstrated [21, 22] that in case of a fragile nanofilament and enough heat delivered to it from a heated cell (HC), the nanofilament can be ruptured due to the cell-to-cell heat transfer, leading to an erasure of a programmed neighboring cell. In some other cases, a combination of the strength of the Cu nanofilament and of the amount of heat transfer leads only to a temporary erasure of the bit, i.e., the ruptured filament recovers spontaneously after a few seconds of a cooling-off period [21, 22]. The objective of this research is to present a thermal methodology that explains the manifold occurrences of electric deterioration induced by thermal transport with reference to the composition, properties of the electrode material, and geometry of the electrodes.

The organization of the chapter is as follows. In Section 2, the fabrication process of our ReRAM devices is described. Various different composite inert electrodes have been fabricated to test the impact of thermal material properties on the thermally

Conductor	ρ_m [g/cm^3]	μ_{th} [W/(mK)]	c_s [J/(gK)]	k_{th} [cm^2/s]	ρ_{el} [Ω cm]
Co	8.9	69	0.419	0.19	6.2×10^{-6}
Cr	7.2	94	0.460	0.28	12.5
Cu	9.0	396	0.395	1.11	1.7×10^{-6}
Pt	21.7	69	0.134	0.24	10.6×10^{-6}
Rh	12.4	150	0.242	0.5	4.3×10^{-6}
Ru	12.4	116	0.239	0.39	7.1×10^{-6}
Ti	4.5	18	0.544	0.07	42.0

Table 1.
Material properties of the metal conductors used in the construction of the inert electrodes: mass density, ρ_m, thermal conductivity μ_{th}, specific heat capacity c_s, thermal diffusivity k_{th}, and electric resistivity ρ_{el}.

Co50 Ti30	Cu150	Pt50 Ti30	Pt50 Cu100	Pt50 Cu200	Rh50 Cr30	Ru50 Ti30	Ru50 Cr30
50	389	50	282	325	129	81	110

Table 2.
Calculated effective thermal conductivity in W/(mK) of the inert composite and active electrodes investigated in this study. The thickness t_i of a material i is given in nanometers (nm). The effective thermal conductivity of two respective materials 1 and 2 is calculated as follows: $\mu_{eff} = (\mu_1 \times t_1 + \mu_2 \times t_2)/(t_1 + t_2)$.

induced electrical degradation of the ReRAM cells. The thermal properties of the materials employed in the construction of the inert electrodes are listed in **Table 1** and effective thermal conductivities are calculated in **Table 2**. The fundamental forming, setting, and resetting I–V characteristics are also presented and explained. In Section 3, we introduce the concept of a "marginal" device as a means to measure the amount of electrical degradation suffered by inactive cells subjected to remote heat transfer from the heated cell. In Section 4, we present thermal analysis based on material properties of the electrode material and its dimensions to predict the amount of degradation of cells subjected to thermal cross-talk. In this section, we show that approximate temperature of the heated cell and of the probed cells can be determined using our thermal analysis based on fundamental thermodynamic mechanisms. We use selected cases of the manufactured composite inert electrodes to verify predictions of our thermal analysis. In Section 5, we summarize the main results of the chapter.

2. ReRAM device fabrication

Resistive random access memory devices are typically arrayed in a crossbar architecture at the junction of two parallel metal electrode lines, as shown in **Figure 1(b)**. The cross-sections of our reference memory cell Cu/TaO_x/Pt and of the ReRAM devices derived from the base line device are illustrated in **Figure 2**. The devices have been manufactured on silicon wafer [3] covered with a silicon dioxide (SiO_2) layer that was 730 nm thick. One observes from **Figure 2** that the inert electrode's design is the only difference between the seven various devices investigated here. Cu(100 nm, 150 nm, and 200 nm), Co(50 nm), Cr(30 nm), Pt(51 nm), Rh(51 nm), and Ti(28 nm) layers have been deposited by physical vapor deposition (PVD) using a Kurt Lesker

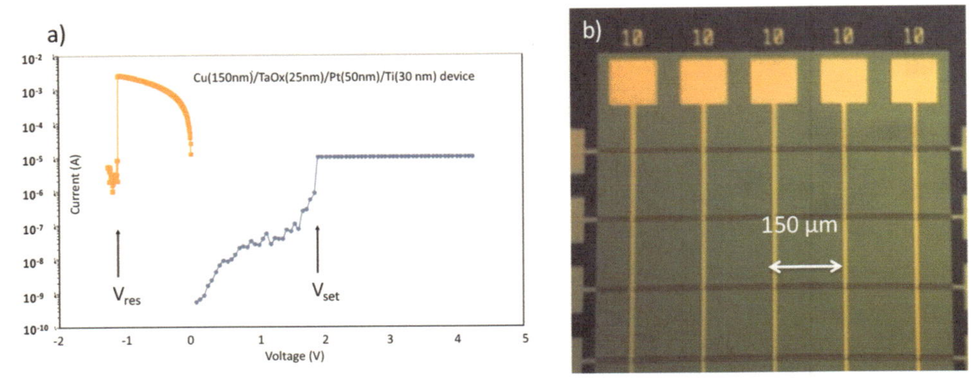

Figure 1.
(a) An example of a SET and RESET operation of the Cu/TaOx/Pt device with I_{cc} = 10 μA. (b) Photomicrograph of the crossline arrangement of the Cu/TaO_x/Pt, Cu/TaO_x/Ru, and Cu/TaO_x/Co devices.

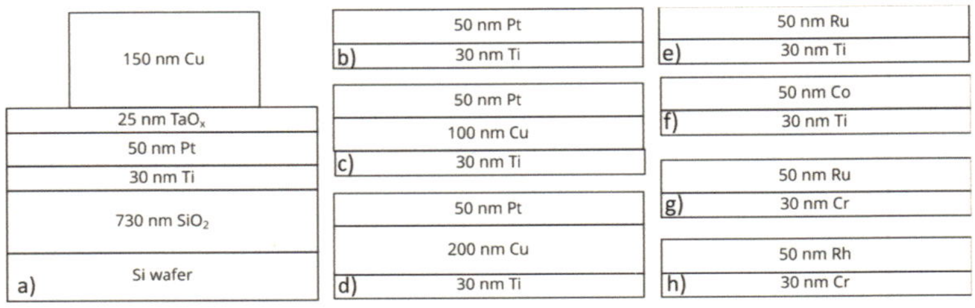

Figure 2.
*Resistive random access memory (ReRAM) arrays with different composite inert (bottom) electrodes that have been assembled, characterized, and analyzed in this work. All devices have the same TaO_x switching film and the same active (top) Cu electrode, 150 nm thick. (a) The figure shows our baseline ReRAM device Cu/TaO_x/Pt/Ti, (b) composite inert Pt/Ti electrode of the baseline device, and (c)–(h) various composite inert electrodes designed to modulate thermal properties of the electrode materials listed in **Table** 2. The thicknesses of the respective layers are also indicated.*

electron-beam physical vapor deposition (EBPVD) equipment, and have been patterned using lift-off technology using photoresist with a thickness of 2 μm. Between Pt, Ru, Rh, Co, and SiO_2, a thin Cr or Ti glue film of 30 nm was utilized to increase the adherence of the inert electrode proper layer and to also modify the thermal characteristics of the inert electrode composed of several materials (composite electrode). TaO_x dielectric has a thickness of 25 nm. To deposit TaO_x film, Ta_2O_5 pellets have been evaporated by electron beam in the evaporation chamber with oxygen-deficient atmosphere, resulting in oxygen-deficient TaO_x ($x \sim 1.9$). The rectangular device regions of the device are between (5 and 35) × (5 and 35) μm^2, depending on the metal line width, which ranges between 5 and 35 μm. The square dimension of the contact pads is 100 μm on the side, and the neighboring line pitch is 150 μm, as indicated in **Figure 1(b)**.

The memory cells employed in this study are thoroughly detailed in [21, 23, 24]. At room temperature, the I–V characteristics were carried out on a probe station with a Keithley 4200-SCS Semiconductor Parameter Analyzer. Keithley equipment lets one select the voltage step height to define the ramp rate, and it chooses the appropriate time step automatically. A ramp rate of 0.276 V/s is a result of a ratio of step height of 0.025 V and a time step of 91 ms, which is internally set by the Keithley analyzer.

Cu^+ ions are produced when a positive bias is provided to the top copper electrode in accordance with the redox reaction, Cu ↔ e^- + Cu^+, and move into the body of TaO_x dielectric in the electric field to be reduced electrochemically at the surface of the Co, Pt, Rh, or Ru counter-electrode, which serves, at the same time, as a reliable diffusion barrier for copper atoms. The copper atom dendrites on the inert electrode expand toward the Cu electrode as more and more Cu atoms gather, and a conductive filament (CF) made of copper atoms acting as building elements creates an electrically conductive channel from the inert Pt to the active Cu electrode. At a threshold voltage V_{set}, when the growing Cu CF makes contact with the Cu electrode, conductance abruptly starts to increase. During the RESET procedure, by applying a negative voltage to the copper electrode while grounding Pt electrode, the filament can be disrupted and the cell's resistance reverts to the HRS. At a threshold voltage, V_{reset}, the filament ruptures abruptly causing an abrupt drop of current in the I–V characteristic.

Figure 1(a) displays common I-V characteristics for RESET and SET operations. To ensure that the cell capacitor is completely depleted, the two grounded probe needles are put on the cell contacts for tens of seconds before any measurement is performed. The voltage of the Cu electrode is then ramped for the SET operation at a ramp rate (RR) in the interval [0.005 V/s to 3.00 V/s]. To prevent damage to the device, a limiting current, the so-called compliance current (I_{cc}) between 5 μA and 1 mA, is applied during the SET operation.

Table 1 lists the key physical properties of the electrode metals used during the manufacture of the devices and their inert electrodes are illustrated in **Figure 2**. In case of composite electrodes, such as the Pt/Cu electrode, the effective thermal conductivity is determined as an average of the weighted respective layer thicknesses that have been deposited. The resistance R_{on} of the LRS can be determined by the application of I_{cc} via equation [2]:

$$R_{on} = C/I_{cc}^{n} \tag{1}$$

where C comes to a value of C = 0.29 V for $n \cong 1$ as extracted from experiments on the Cu/TaO_x/Pt devices. When n is unity, the constant C signifies the lowest voltage possible under which the device can be switched to the ON state, as demonstrated in [2]. This has been verified by delivering a constant bias to a cell to reset it while observing the time dependence of the current [2]. The cell may be switched to a conductive or ON-state for any voltage exceeding 0.286 V, but for voltages less than 0.286 V, the cell remains in the nonconductive state, i.e., OFF-state, even when the constant voltage is maintained for an extended time period. The relationship between R_{on} and I_{cc} enables the formation of both strong, low resistive, R_{on} = 500 Ω, at I_{cc} = 0.2 mA, and weak, highly resistive, R_{on} = 50kΩ, Cu filaments at I_{cc} = 5 μA. The excellent retention and endurance of the Cu/TaO_x/Pt-type devices have been extensively discussed in [25].

The filament either does not develop at all for compliance current values below 5 μA or it becomes volatile and ruptures spontaneously without being stressed. Contrarily, cells with strong low-resistance Cu filaments are more likely to become inoperable when I_{cc} > 0.25 mA, indicating that the device has been permanently damaged and became nonresettable.

In this study, we are interested in the study of cells that, as shown in **Figure 3**, are connected to the heated device by a thermal conductance channel and have been placed into a LRS condition under predetermined levels of compliance current, I_{cc}, before the transfer of heat from the heated cell (HC) to the probed cell is being applied.

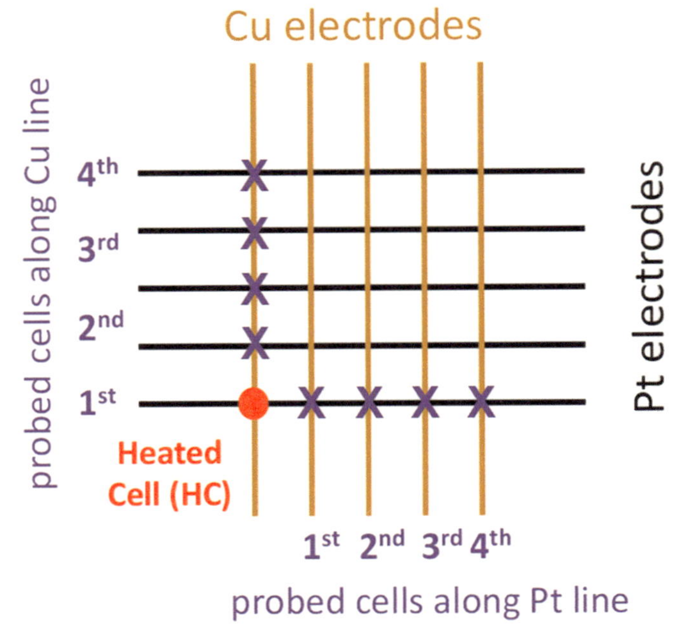

Figure 3.
A schematic resistive random access memory (ReRAM) crossbar array shown for the baseline device. The cell marked by the red dot represents the heated cell (HC). The neighboring devices impacted by the heated cell are positioned alongside the Pt and Cu metal lines and marked by blue crosses. The neighboring probed cells are set prior to heating by the cell HC to an ON-state. Immediately after the heating by the HC cell comes to an end, the neighboring cells are checked to determine the corresponding maximum switching cycle (SWC).

A repeated switching of a device results in the dissipation of Joules heat in that device, which can disperse to nearby devices placed along one of the metal lines common with the heated device, as further explained in [8, 20]. As long as the other array cells are switched to the conductive state, where the Cu nanofilament acts as a conductor for the heat transfer, even those devices that do not share any of the two electrode lines with the heated device may be impacted by the cell-to-cell heat transfer as well. Depending on the heat generated in the heated device, the kind of common electrode selected (Cu or Pt), and the distance between the probed and the heated device, this phenomenon permits to heat a neighboring cell in a controlled gradual way (see **Figure 3**).

The maximum temperature of the filament, according to Sun et al. [26], lies between 600°C and 900°C. The majority of the models published in the literature consider the mesoscopic electrodes to be a constant heat sink condition fixed at ambient temperature. Our findings completely disprove this presumption. Heat transfer between the array's cells is made possible by the electrode lines' significant heating. Joules heat in metallic nanofilaments with various wire cross-sections has been investigated by Fangohr et al. [27] who discovered that the temperature at the constriction can reach the high temperature of 1336 K. By using infrared (IR) detection, Uenuma et al. [28] observed a hot spot on the surface of the top electrode indicating the position of the nanofilament and estimated the temperature of the spot to lie well above 900°C. Sato et al. [29] calculated that the temperature of filaments with radii ranging between 10 nm and 40 nm will result in a 1000–1200°C temperature interval, depending on the filament's resistance. There exists, therefore, broad agreement that when the device is switched back and forth at a high frequency of

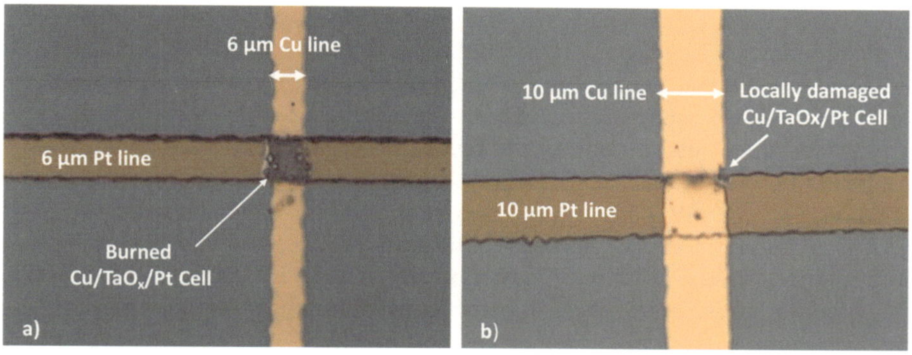

Figure 4.
Damage to the Cu/TaO$_x$/Pt cells due to excessive heating. Both cells have been set and reset under the same conditions (i.e., the same voltage ramp rate and the same I_{cc} during the SET operation). (a) The cell on 6 µm Cu and Pt lines shows complete melting of Cu over the cell after three RESET-SET cycles. (b) The cell on 10 µm Cu and Pt lines shows local melting of Cu along the Pt lines after six RESET-SET cycles. After the damage is shown, both cells became inoperable.

SET-RESET cycles, the maximum temperature of the filament can readily surpass 900°C. In Ref. [30], it has been experimentally observed that frequent switching of a ReRAM cell leads to the incipient melting of the Cu electrodes including Cu pads. This gives a definite reference point for the local temperature of 1085°C, which is the melting temperature of copper. In **Figure 4(a)**, it can be seen that the entire segment of the Cu electrode of a 6 µm wide device has been removed after three RESET cycles performed under the same bias conditions. **Figure 4(b)** shows the result of an irretrievably damaged cell after six identical RESET cycles as before for a 10 µm Cu and Pt line. We will discuss the impact of the line width on the cell temperature and severity of melting in Section 4 in more detail. Here, we wish to point out that from **Figure 4(b)** it can be seen that the location of local melting is visible on the perimeter of the cell along the Pt electrode's edge, a preferable location of the nanofilament, where the TaO_x coverage is slightly less than on the planar surfaces, which is an intrinsic property of limited conformality of the EBPVD deposition. The Cu electrode above the cell is damaged by electromigration, which is triggered by current crowding at the nm-sized region where the Cu nanofilament makes contact with the Cu electrode. As a result, electromigration depletes the Cu atoms at this interface. The extremely high current densities and Joule heating that raise temperatures to the melting point of copper exacerbate this phenomenon, which is similar to the effects of electromigration in vias and contact holes.

3. Effects of thermal cell-to-cell cross-talk in ReRAM arrays

N RESET-SET cycles were applied to heat the cell incrementally, in well-defined quanta of heat. It should be observed that compared to the RESET cycle, the heat dissipation during a SET cycle is extremely low and negligible. At the beginning of the voltage ramp, the current is less than 1 nA during the SET operation until V_{set} is reached, and then does not exceed the I_{cc} limit of less than 100 µA when threshold voltage V_{set} is surpassed. However, the current flowing through the device, when it is reset, is in the range of a few milliamperes (mA), i.e., at least more than two orders of magnitude higher. As a result, heating primarily occurs during RESET operations, and, therefore, resembles a triangular-shaped heat pulse with a pulse duration of several seconds given by $t_{res} = V_{res}/RR$ and pulse periods of 40–60 s, which correspond to the time needed to reposition the probe needles of the probe station between the application of changing bias conditions from RESET to SET. This indicates that when the cell gets heated up during the RESET operation, there follows a relatively lengthy cool-off period of about 40 s. As a result, the heating's duty cycle is less than about 0.05. The cell and the entire cell array will partially cool during the off-pulse period as a result of attending Newtonian heat losses. Only a portion of the heat deposited is kept in the cell, and this portion is described by the factor f_{loss}. As a consequence, it takes several cycles for the device to build up enough heat to achieve a target temperature in the cell. The number (N) of RESET-SET cycles, the magnitude of the voltage ramp rate (RR) applied, and the ON-resistance, R_{on} resistance of the nanofilament, can all be employed to modulate the amount of Joule heating generated at the heated memory device.

The electrical characteristics of a virgin cell have been found to be impaired when a neighboring device has been heated by the application of multiple RESET-SET cycles. This was found when all memory devices joined by a thermal path to the heated device were electrically characterized later. It has been observed that the electrical reliability of the neighboring device degrades significantly when subjected to this kind of remote heating. However, when the affected probed neighboring cell was examined again after 3–5 min or a longer cool-off time, this deterioration of electrical characteristics disappeared. We assume that after a period of time in the order of 2–5 min, the thermally induced degradation abates gradually. During the electric testing of the neighboring cells, we made sure beforehand that there exist no sneak pathways [31].

To measure the impact of the cell-to-cell heat transfer talk [19, 20] on neighboring cells, we have established the concept of a "marginal" memory cell as a convenient metric. Cell's property of displaying only a very small number of RESET-SET cycles is what we mean when we say that a cell is "marginal": We have configured a neighboring cell with an I_{cc} of just 10 μA and a voltage ramp rate of RR = 1.1 V/s to form a weak filament in the cell.

After specific number of consecutive RESET-SET switching cycles, such cell set at I_{cc} = 10 μA turns volatile. RESET events that occur on their own, i.e., without any external bias, are one example of cell's volatility. When the cell's on-resistance R_{on} is adjusted at higher and lower I_{cc}, the nature of marginality of the memory cell is further elucidated. The device cannot hold the LRS when I_{cc} = 7 μA is imposed on the SET operation. A nanofilament set at such low I_{cc} ruptures spontaneously. The ON-state remains stable when the device is placed at higher level of I_{cc}, such as 10 μA, but even here only for a very limited number of consecutive SWC, usually 11–13, depending on the specific cell. When this number of SWC is reached, the cell's electrical performance turns erratic. To firm up this metric, 100 cells with an I_{cc} of 10 μA were tested, and the maximum number, M_x, of consecutive switching RESET-SET cycles was registered. We find that average maximum number of SWC is M_x = 12.7 cycles, with a standard deviation of =1.3 for such "marginal" cells.

To allow for a lengthy heating period during the RESET procedure, a low RR = 0.1 V/s has been selected. Because at low RR, the high reset current persists for a prolonged period of time, low RR allows for large deposition of heat in the cell. It has been demonstrated in [32] that the reset voltage V_{res} decreases as the voltage ramp rate, RR, decreases, according to $V_{res} \sim \sqrt[3]{RR}$. When the maximum number of SWC, (M_x), for the cell has been reached, the cell's performance becomes extremely unstable.

In contrast to the "marginal" cell, a cell set into the ON-state at high I_{cc} = 40–100 μA is very stable and may be switched very frequently (hundreds of times and more). In summary, a "marginal" memory cell marks the boundary line between the cell's volatile and stable switching performance. As a result, our "marginal' device with a filament formed at I_{cc} value of 10 μA, and allowing for at most M_x, of RESET-SET SWC, serves as the indicator for determining the severity of the cell-to-cell heat transfer.

It has been found consistently that such a marginal cell is very sensitive to the amount of heat being transferred between the cells. Any significant heat transfer is bound to reduce the maximum number M_x = 13 to a lower value, which serves then as the indicator of the degradation of electrical performance. As a result, the onset of the device's instability in response to its local temperature is mirrored in the metric M_x. As a consequence of the heating by the neighboring cell, the heat dissipated locally by the cell's intrinsic switching is building on the heat provided by a remote source. The time it takes to change the contact needles on the probe station is approximately 50 s in our experimental setup between the point at which the heat cell's heating is stopped and the probed cell's characterization begins.

The M_x for a remotely preheated cell is compared to the M_x of the same cell in the absence of remote heating, i.e., unstressed cell, to establish then the degradation metric defined as follows:

$$DEG = \frac{M_x(unstressed\ cell) - M_x(preheated\ cell)}{M_x(unstressed\ cell)} \tag{2}$$

As previously indicated, it has been discovered that the average M_x for a cell without remote heating (unstressed cell) is about 13 RESET-SET cycles. Based on 125 devices evaluated, the results of cell's deterioration of switching ability for devices positioned along the Cu and Pt electrode lines are displayed in **Table 3** for four different devices with different inert electrodes. The maximum SWC are then used to determine the degree of degradation suffered by the cells due to the remote heating.

As seen in **Table 3**, for the device 1 (our baseline and reference device), Cu (150 nm)/TaO_x/Pt(50 nm)/Ti(30 nm), 10 μm wide line, the deterioration of SWC for the first neighbor along the Pt line is approximately DEG = 67%. The fourth device, also attached to the Pt electrode, has a substantially reduced rate of degradation (DEG = 13%). The amount of thermal cross-talk depends on the distance in units of 150 μm between the heated and probed cells, or n × 150 μm, where n = 1, 2, 3, and 4 stand for the first, second, third, and fourth adjacent cells, respectively, as shown in **Figure 3**.

As seen from **Table 3**, for the device 1, the degradation of the adjoining cells positioned alongside the Pt electrode is far less severe than that of the corresponding cells disposed alongside the Cu metal line, as extensively detailed in [20]. For example, the first neighbor along the Cu line to the heated cell suffers a degradation of 80% compared to 67% for first neighbor alongside the Pt line. At first, this result was unexpected to learn that the deterioration along the platinum metal line is less severe than that along the copper metal line [20], as one could anticipate that the geometry of the filament, due to its genesis, should have a larger contact footprint with the Pt electrode and a small contact footprint of the top part of the Cu CF with the copper electrode. Because of the above observation, one is forced to revise the hypothesized configuration of the filament from a truncated cone to a something resembling more

1. Cu(150 nm)/TaO_x/Pt(50 nm)/Ti(30 nm) Next neighbors to the heat cell along Cu(150 nm) 10 μm wide line					1. Cu(150 nm)/TaO_x/Pt(50 nm)/Ti(30 nm) Next neighbors to the heat cell along Pt(50 nm) 10 μm wide line			
prob. Cell #	I	II	III	IV	I	II	III	IV
M_x	2.6	3.3	3.9	4.3	4.8	6.1	7.8	11.3
DEG [%]	80	75	70	67	67	53	40	13
2. Cu(150 nm)/TaO_x/Pt(50 nm)/Cu(100 nm) Next neighbors to the heat cell along Cu(150 nm) 10 μm wide line					2. Cu(150 nm)/TaO_x/Pt(50 nm)/Cu(100 nm) Next neighbors to the heat cell along Pt(50 nm)/Cu (100 nm) 10 μm wide line			
prob. Cell #	I	II	III	IV	I	II	III	IV
DEG [%]	57	41	0	0	23	1	0	0
3. Cu(150 nm)/TaO_x/Pt(50 nm)/Cu(200 nm) Next neighbors to the heat cell along Cu(150 nm) 10 μm wide line					3. Cu(150 nm)/TaO_x/Pt(50 nm)/Cu(200 nm) Next neighbors to the heat cell along Pt(50 nm)/Cu (200 nm) 10 μm wide line			
prob. Cell #	I	II	III	IV	I	II	III	IV
DEG [%]	11	0	0	0	0	0	0	0
4. Cu(150 nm)/TaO_x/Pt(50 nm)/Ti(30 nm) Next neighbors to the heat cell along Cu(150 nm) 35 μm wide line					4. Cu(150 nm)/TaO_x/Pt(50 nm)/Ti(30 nm) Next neighbors to the heat cell along Pt(50 nm) 35 μm wide line			
prob. cell	I	II	III	IV	I	II	III	IV
DEG [%]	65	52	40	33	43	31	19	0

Table 3.
The average maximum number of RESET-SET switching cycles (SWC) and the amount of electrical degradation of neighboring devices alongside Cu and Pt electrode lines exposed to the remote heating by the heated cell for four different devices with different inert composite electrodes: 1. Pt(50 nm)/Ti(30 nm), 10 μm wide line, 2. Pt (50 nm)/Cu(100 nm), 10 μm wide line, 3. Pt(50 nm)/Cu(200 nm), 10 μm wide line, and 4. Pt(50 nm)/Ti (30 nm), 35 μm wide line.

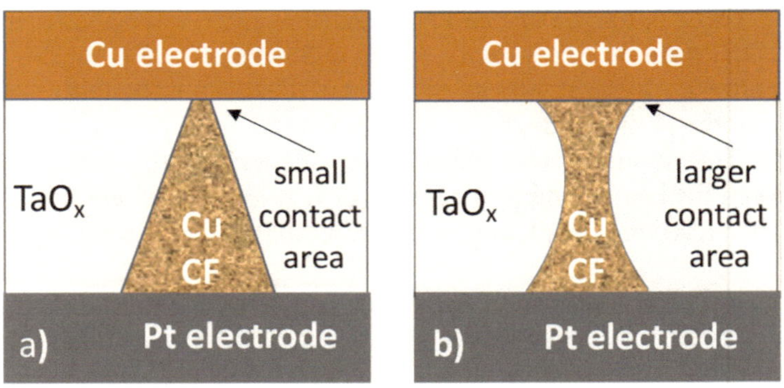

Figure 5.
Hypothesized geometrical shapes of the Cu conductive nanofilament. (a) A truncated cone conventionally assumed with a small contact area with the Cu electrode. (b) An hour glass silhouette providing larger contact area with the Cu electrode.

an hour glass silhouette, which would exhibit similar size contact areas with Cu and Pt metal lines as shown in **Figure 5**. Given that it now has a larger contact area with the Cu electrode, the hourglass-shaped filament would more accurately represent our data. Consequently, it would be easier to understand why there is such a significant heat transfer from the filament along the Cu electrode line to the nearby cells along the Cu electrode. Apparently, the substantially higher cell-to-cell heat transport along the copper electrode than alongside the platinum metal line is the root cause for the more severe deterioration in electrical performance of the cells positioned along the copper metal line.

From the heat conduction transport mechanisms, it is known that the product of the electrode's cross-section and its thermal conductivity determines the rate of heat transmission. The rectangular cross-section of the Pt/Ti electrode is two-and-one-half times smaller than that of the copper electrode, and the thermal resistivity of copper is about eight times smaller than that of weighted thermal resistivity of the composite Pt/Ti (see **Table 2**). As a result, Cu electrode's rate of heat transmission is in excess of 20 times that for the heat transfer rate for the platinum electrode. Hence, the adjacent cells positioned on the Cu electrode reach higher temperatures than the corresponding cells disposed along the platinum metal line.

4. Prediction and verification of ReRAM devices' electric degradation

The amount of I_{cc}, the number of consecutive RESET-SET SWC, N, and the rate at which the voltage ramps up during the RESET cycles, RR, can all affect how hot the heated cell (HC) (see **Figure 3**) becomes. For the RESET operation, a low RR = 0.1 V/s and/or a high I_{cc} that results in a strong low resistance Cu CF may be chosen to enhance heating by allowing for a lengthy heating period before the threshold voltage V_{res} is reached. Low ramp rate maximizes the cell heating because the current persists for a long time during the RESET operation until it reaches the critical value I_{res}. Eq. (3) [32, 33] can be used to compute the Joules heat (JH) dissipated in a device when an increasing voltage bias with a constant voltage ramp rate is applied, and where the relationship between R_{on} vs. I_{cc} given in Eq. (1) is used to derive the final expression for JH.

$$JH = \int_0^{t_{res}} \frac{V^2(t)}{R_{on}} dt = \int_0^{V_{res}/RR} \frac{RR^2 \times t^2}{R_{on}} dt = \frac{V_{res}^3 \times I_{cc}}{3 \times RR \times C} \tag{3}$$

Here, $t_{res} = V_{res}/RR$ [32, 33] determines the time of voltage ramp rate application to effect the rupture. JH can range from 3 to 60 μJ depending on the I_{cc} and RR parameters that are used. The experimental values for R_{on} and V_{res}, and the applied voltage ramp rate, RR, solely determine the amount of JH, which is the energy that is being dissipated in the memory cell during the RESET action. The nanofilament, which is characterized by the ON-resistance R_{on}, and completely embedded in the dielectric, is the main source of heat generation. How the generated heat is being stored is a concern to which we advert now. Material's temperature as a result of the heat transfer rises according to Eq. (4):

$$Q = \Delta T \times c_s \times m \tag{4}$$

where the total mass $m = V \times \rho_m$. The energy that was dissipated within the memory array, JH, is determined by the experimental values for R_{on} and V_{res} and by the applied voltage ramp rate (RR). Eq. (4) shows, among other things, that the volume of the electrode affects how much the temperature difference ΔT increases.

To calculate ΔT, one must determine the total mass of the electrode, which is determined by the product of the mass density ρ_m of the electrode material and by the electrode's total volume V ($V = width \times length \times thickness$) including also the contact pads' volume at both ends of the electrode. Evidently, the electrode's temperature increase ΔT is inversely correlated with both the thickness and the width of the electrode. As a result, the geometry, or the electrode's volume, controls how much the temperature may rise by ΔT. For the time being, let us suppose that the entire amount of heat JH produced by the filament is held in the Cu conductive filament. The question of what temperature difference ΔT would be required to hold a heat of 10 μJ in a Cu CF then arises. It requires estimation of the mass density, specific heat, and volume of the Cu NF in order to be answered. The volume of the nanofilament equals 4.6×10^{-18} cm^3 if we assume that the nanofilament has the shape of a truncated cone with a radius of 9 nm at the base, 4.5 nm at the top, and 25 nm height (a calculation for an hourglass-shaped CF would be similar), which is limited by the thickness of TaO_x [34]. If the electrical resistivity of the Cu CF is assumed to be $\rho_{el} = 5\times10^{-4}$ Ω cm, which is around 20 times greater than that of bulk Cu, then this volume results in a R_{on} resistance of about 1000 Ω. At I_{cc} = 10 μA, such a R_{on} value is in fact experimentally observed. Here, the resistivity of the Cu filament has to be larger than that of bulk copper, since CF consists of Cu atoms embedded in the matrix of TaO_x. Given that Ta_2O_5 has a mass density of 8.2 g/cm^3 and that of Cu is 9.0 g/cm^3, the filament's mass density must fall between the two values. If it does, then the total mass of Cu CF is about 4×10^{-17} g. Furthermore, the specific heat capacity of copper filament must fall between that of Ta_2O_5 of about 750 J/(gK) and Cu's 0.385 J/(gK). One may determine the temperature differential ΔT required to store such energy within the nanofilament alone using Eq. (4) and the extreme case of 750 J/(gK). Under these circumstances, the temperature differential from the ambient temperature is at least 3×10^8 K, which is 20-fold higher than the core temperature of the sun of 1.5×10^7 K. The temperature would be even higher (3×10^{11} K) if the specific heat capacity was at its other extreme, c_s (Cu-CF) = 0.385 J/(gK). This is obviously impossible, and given the filament's incredibly small volume, it is evident that this amount of energy cannot be accommodated by the filament. Instead, the heat must be partially dissipated and partially stored, through the primary conductive heat transfer mechanism, at least in the two electrodes (Cu and Pt), and in the body of the nanofilament with its negligible contribution to the overall volume. Additionally, a significant portion of the energy conductively transferred to the electrodes will be lost owing to Newtonian heat losses, such as buoyancy-driven convection and radiation, all the more as the metal lines display a relatively large surface-to-volume ratio [35]. Therefore, it is reasonable to suppose that only a small portion of JH will be accommodated in the body of the electrode over the course of a more or less continuous heating session during the reset procedure, or $Q_{st} = JH \times f_{loss}$, where f_{loss} signifies a portion of the heat still present in the material after the losses of heat to the environment. The relationship between the coefficient f_{loss} and the ambient temperature is $f_{loss} \approx \exp.(-rt_{diss})$, where r denotes the coefficient of heat transfer and t_{diss} the dissipation time [36]. The remnant heat, Q_{Nst}, after considering the losses to the environment (f_{loss}) and amount of cooling off (f_{diss}) between the switching operations stored in the electrodes when a memory cell is quickly switched to and from N times is given then by Eq. (5).

$$Q_{Nst} = N \times Q_{st} \times f_{diss} = N \times Q_{JH} \times f_{loss} \times f_{diss} = c_s \times m \times \Delta T \qquad (5)$$

where f_{diss} is a coefficient smaller than unity and reflects the heat dissipation time between SWC and accounts for both the heat loss that occurs when heat is transferred between heated and probed neighboring cells during the set operation as well as the cooling period between the set operation when no substantial current flows until V_{set} is reached or very little current flows when the bias exceeds V_{set} but is still limited by I_{cc}. The mass of the two electrodes joined by the filament is now reflected in the mass parameter in Eqs. (4) and (5). To make practical use of Eq. (5), we avail ourselves of a valuable experimental reference [30] that can be used to calibrate Eq. (5). When a cell is repeatedly switched on and off under specific switching conditions for moderately large R_{on} and low value for RR, which were purposefully designed to increase the amount of heating, we witness melting of the copper electrode at the heated device that extends to the metal line pads [30]. It has also been noticed that the contact pads for the Pt electrodes are free of any visible distortion of their square shape on account of much higher melting temperature of Pt (1768°C) than that of Cu (1085°C). The reflowing of the Cu electrode has been observed after 25 SWC for a cell whose heat dissipation according to Eq. (3) is JH = 40 μJ based on the values utilized for RR, as well as experimental values for R_{on} and V_{res}. Using c_s = 0.385 J/(gK), for Cu and c_s = 0.13 J/(g°K) for Pt, respectively, we can compute the product $f_{loss} \times f_{diss}$ from Eq. (5) and arrive at $f_{loss} \times f_{diss}$ = 0.063.

With this knowledge, it is now possible to determine the average temperature rise in the CF-electrode system utilized in the tests described in **Table 1** after a single switching cycle, where one reset cycle has JH = 10 μJ. Since the contact pads are square in shape with 100 nm on the side, we can repeat the calculation using Eq. (4) and determine the T for the specific heat capacities of the Cu and Pt electrodes, c_s = 0.385 K/J/(gK) and c_s = 0.13 J/(gK), respectively. Using the heat released in one switching (reset) cycle of JH = 10 μJ and assuming tentatively f_{loss} = 0.1, one can infer the temperature increase is ΔT = 27°C. Obviously, the temperature at first is not uniform along the electrode lines when the heated device begins to heat up, but as was already mentioned, due to the high thermal diffusivity, it quickly reaches the steady state. The highest temperature of the heated device will be transmitted to the adjacent cells resulting in lower temperature of the neighboring cells. For an isolated metal rod, the 1-D heat equation's solution can be used to accurately predict f_{diss} [37]:

$$\frac{\partial T}{\partial t} = k_{th}\frac{\partial^2 T}{\partial^2 x} + \frac{g}{c_s \rho_m} \tag{6}$$

where g is the heat produced, k_{th} is the thermal diffusivity, c_s represents the specific heat capacity, T denotes temperature, and ρ_m stands for the mass density. The explicit solution to Eq. (6) for the metallic rod with a heated end can be found in [37]. It should be observed that TaO_x has a thermal diffusivity of k_{th} = 0.02 cm^2/s, while the thermal diffusivities of Pt and Cu are much larger (see **Table 1**), thus approximating well the isolated rod model. The time defined by $t_{ts} = L^2/k_{th}$ provides the typical time scale for Eq. (6). With a cell-to-cell distance L of 150 μm and a thermal diffusivity of Cu k_{th} = 1.11 cm^2/s, one can calculate t_{ts} to be ~200 μs. In view of this, one is justified to evaluate Eq. (6) under steady-state conditions because the period of the pulse of the SET-RESET cycles is about 60 s, which is 3×10^5 times larger than t_{ts}. Under steady state, the temperature becomes a linear function of the spatial coordinate alongside the electrode. Surely, during the 60 s of idle time, the electrode will cool down as a result of Newtonian heat losses, and the amount of heat still stored in the electrode can be approximated using the parameter f_{loss} as previously explained. Thus, if a corner array cell is heated up, the temperature will be a linear function of a distance along both of the electrodes.

One can calculate the transit time, t_{tr}, needed for a given temperature value to travel the distance of D = 850 nm, from the heat device to the pad on the far end of the metal line using the equation $D = \sqrt{k_{th} \times t_{tr}}$. This duration is less than 10 μs for the copper electrode and 2.23 times larger for the platinum metal line on account of the latter's fivefold lower thermal diffusivity (see **Table 1**) as well as the smaller cross-section of the Pt metal line than the Cu line. This makes it plain that, for the neighboring devices being studied after the heated cell's heating has been terminated for 45–65 s, the temperature distribution is well approximated by a steady state. The result is a linear dependence of the temperature on the spatial coordinate designating the location of the neighboring cells. Considering that the first cell positioned on the Cu or Pt electrode is heated up to temperature T_c while the electrode's farthest end is kept at temperature $T_o < T_c$, the temperature decrease from device to device is given by $(T_c - T_o) \times 150/(100 + 5 \times 150) = (T_c - T_o) \times 0.18$, where 150 μm is the distance between two immediate neighboring cells and $100 + 5 \times 150 = 850$ μm is the total length between the heated device and the contact pad at the far end of the line. T_o at the very beginning of the heating will be still 27°C but it will rise as the heating keeps persisting. When the heat loss from the neighboring cells that are the farthest away is taken into account, a reasonable estimate of f_{diss} is between 0.6 and 0.7. The estimate of f_{diss} enables us to estimate f_{loss} to be between 0.11 and 0.09, which renders our aforementioned choice of $f_{loss} = 0.1$ self-consistent. This determination is based on the experimentally extracted value $f_{loss} \times f_{diss} = 0.063$. As a result, heated cell's threshold temperature T_c decreases as heat is transmitted to the adjacent cells via one of the common electrodes. The robustness of the Cu CF determines the threshold temperature, T_c for the specific cell, at which the cell's instability sets in, and as a result, T_c is bound to increase with the increasing strength of the filament or its decreasing R_{on} resistance. The critical cell temperature for a weak filament with a R_{on} resistance value of several tens of kiloohms (kΩ) has been determined to be around 350°C.

The number of SWC precipitating a rupturing of a Cu filament of a specific strength can be used to estimate the temperature increase ΔT for a single RESET operation. We found previously that it took 25 cycles of heating of a heated cell with a robust Cu filament of R_{on} = 5kΩ set at I_{cc} = 100 μA to see the beginning of the reflowing of the contours of the Cu electrode pads, indicating an incipient melting of the copper metal. As a result, we may estimate ΔT for a single cycle using the formula ΔT = (1085°C – 27°C)/25 = 42°C. The experiment was then performed with the same probed and heated cells, except that the probed cell was initially set with a thin filament of high resistance R_{on} = 20 kΩ formed at I_{cc} = 20 μA, and we found that 7–8 heating cycles have been enough to cause the device to rupture. The temperature is raised by 42°C in one cycle, hence 42°C × 8 + 25°C = 361°C. Thus, a value of around 350°C represents a crucial temperature for rupturing the weaker Cu filament. However, a robust copper nanofilament with low resistance of a few hundred Ohms is likely to display a critical temperature needed to rupture it near the Cu melting temperature, i.e., 1085°C. The threshold temperatures T_i for a weak filament for each of the five cells arranged along the Pt line are plotted in **Figure 6**, with the first cell serving as the heated device and the other four serving as remotely heated neighboring cells. The heated device stressed with 13 maximum RESET-SET cycles reaches the instability threshold of T_c = 350°C. The nearby cells will exhibit lower temperatures, $T_c \times f_{diss}$, $T_c \times f_{diss}^2$, $T_c \times f_{diss}^3$, and $T_c \times f_{diss}^4$, for the first, second, third, and fourth neighbors, respectively, as explained above.

In **Figure 6**, the temperature difference $\Delta T = T_c - T_i$ between each neighboring cell and T_c is shown. This temperature margin ΔT limits how many inherent heating cycles the cell can withstand before the cells becomes unstable. For the first neighboring cell, this difference is 130°C. Since each heating cycle increases the temperature effectively by 27°C, the cell can switch maximally for 130°C/27°C = 4.8 or 5 cycles, while the fourth neighboring cell has a temperature margin of (350 – 50°C) = 300°C, which allows for maximum 300°C/27°C = 11.1 11 cycles. Using our definition of degradation given in Eq. (2), these two cases would correspond to degradation of the first cell of (13 – 5)/13 = 62% for the first neighboring cell and (13 – 11)/13 = 15%, respectively, in perfect agreement with the experimentally measured values shown in **Table 3**.

The maximum SWC M_x can be computed in the same way for all other cells. The corresponding M_x numbers are also indicated in **Figure 6**. Thus, there is a striking match between the expected number of the extra permissible RESET-SET cycles and the respective experimental figures M_x(ave) provided in **Table 3**. It should be noted that the temperatures shown in **Figure 6** represent a snapshot of a transient phenomenon when the peak temperatures have been reached just after the heating of the heat cell has been stopped. If the cells are then probed just afterwards, they display degraded switching properties in terms of reduced number of switching cycles, as evidenced in **Table 3**. The transient nature of this phenomenon precludes, however, the determination of the endurance and retention properties.

We have tested the predictions based on the methodology described above by building ReRAM arrays with modified thermal properties of the electrodes. To

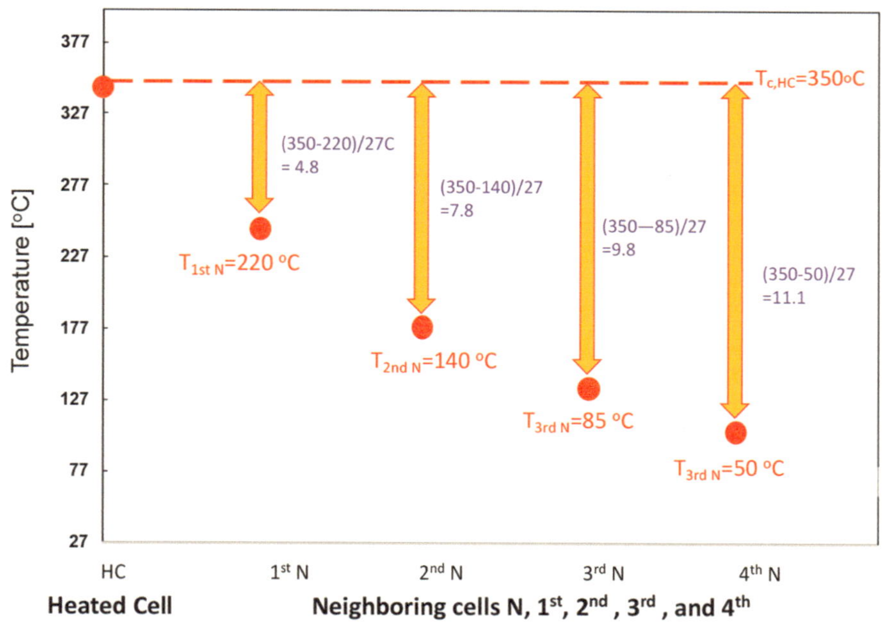

Figure 6.
The red dots show the temperature of the heated cell when it reached a threshold temperature T_c of 350°C as well as of the four neighboring cells. The yellow double arrows indicate the temperature difference to the threshold temperature, which determines the amount of the cell's intrinsic heating to reach the threshold temperature at which the cell becomes unstable. Since one heating cycle increases the cell temperature by 27°C, the ratio of $(T_c - T_{i\ N})/27$ (where i = 1, 2, 3, 4) determines how many heating cycles the cell can tolerate before becoming unstable. For example, the fourth cell can tolerate 11 cycles before it becomes unstable, in very good agreement with the experimental degradation data.

accomplish it, we have manufactured, in addition to the Pt(50 nm) inert electrode, two composite Pt(50 nm)/Cu(100 nm) and Pt(50 nm)/Cu(200 nm) electrodes, as well as Rh(50 nm). It should be noted that for composite Pt/Cu electrodes, the intrinsic layers defining the memory cell are the same as for the Pt cell. The additional Cu layers present just two different embedment scenarios external to the cell proper, but are of crucial importance for the heat conduction properties. The key material characteristics of Cu, Pt, and Rh are listed in **Table 1** and the effective thermal conductivities are calculated in **Table 2**. For the addition of Cu layers of 200 nm and 100 nm, respectively, the thermal conductivity of the bottom inert electrode increases 4 times, and the product $V \times c_s = m$ increases 1.95 and 1.47 times, respectively.

We now apply this methodology and compute the temperature for a single heating cycle by repeating the similar I–V measurements of the deteriorated cells applying the same RESET parameters used for calculating JH in Eq. (3). A temperature increment of 26.5°C for the Pt(50 nm) array, 18.1°C for Pt(50 nm)/Cu(100 nm), and 13.5°C for Pt(50 nm)/Cu(200 nm) has been measured, respectively, for one SWC. It is intuitively obvious that with the bottom electrode's increased ability to store and conduct heat, it will take more heating cycles to reach the same threshold temperature T_c upon reaching which the device is rendered unstable. Because of the bottom electrode's substantially better thermal conductivity, which causes a much faster transport of the heat pulse across the length of electrode, the maximum number of SWC is bound to increase.

The thermal diffusivity increased fivefold as a result of the composite inert electrode's enhanced thermal conductivity (Pt(50 nm)/Cu(200 nm), and the coefficient f_{diss} for the Pt(50 nm) should have been at least three to four times smaller. This would suggest that the maximum number of switching steps for the ReRAM array with modified inert electrode with the same switching parameters should be 6–8 times greater, or 75–100 maximum SWC. The corresponding metrics (ΔT and M_x) for an array with Pt(50 nm)/Cu(100 nm) are $\Delta T = 18.0$°C and a maximum RESET-SET cycle of approximately $M_x = 25$. **Table 4** displays the information on the maximum RESET-SET cycles for the Rh electrode, the Pt/Cu(100 nm), and Pt/Cu(200 nm) inert electrodes.

The results of similar electrical degradation tests executed on the Pt(50 nm)/Cu (100 nm) and Pt(50 nm)/Cu(200 nm) ReRAM memory crossbar arrays in the same way as tests performed on Pt(50 nm) arrays are shown in **Table 3** for devices #2 and #3, respectively. It is seen that the cells positioned along the Pt electrode do not degrade at all in the case of the Pt(50 nm)/Cu(200 nm) (device #3) inert electrode, while the first neighboring device positioned alongside the Cu electrode degrades only

Inert electr.	Pt(50 nm)/Cu(200 nm)	Pt(50 nm)/Cu(100 nm)	Rh(50 nm)	Pt(50 nm)
M_x	75	25	26	13
μ_{eff} [W/mK]	325	282	79	50

Table 4.
The measured maximum M_x of RESET-SET SWC for a fragile filament set at the same value of I_{cc} for four different inert electrodes. The last column in the table shows our baseline device with the Pt(50 nm)/Ti(30 nm) with the maximum mx of 13, as described already in Section 3. The effective thermal conductivity of the inert electrodes is also displayed to demonstrate palpably the clear correlation of maximum SW with the effective thermal conductivity of the inert electrode.

by 11% while the subsequent neighbors do not degrade at all. In contrast to the Pt (50 nm)/Ti(30 nm) ReRAM array (device #1), the local temperature for Pt(50 nm)/ Cu(200 nm) is clearly considerably lower due to the better thermal conductivity and specific heat capacity of the Pt(50 nm)/Cu(200 nm) electrode. Since fresh cells fail for a Pt(50 nm) electrode after just 13 times, but now for 72–90 times for a Pt(50 nm)/Cu (200 nm), this type of experimental testing undertaken for this study is rather laborious, but has been undertaken to demonstrate the accuracy of the predictions of our thermal analysis. As one would expect, the degradation data shown in **Table 3** for the inert electrode Pt(50 nm)/Cu(100 nm) (device #2) lie between those of the Pt(50 nm) (device #1) and Pt(50 nm/Cu(200 nm) (device #3).

The suggested thermal study can be expanded to include the relationship between cell degradation and the width of the electrode lines. According to Eq. (4), the volume of the electrode grows proportionally with the width, causing the temperature difference ΔT to decrease with wider widths of the electrode lines. Pt and Cu lines (10 μm) have served as the standard for all of our tests thus far. We have gone over the analogous study for the 35 μm wide Pt line in the Pt(50 nm) ReRAM array. The mass of the platinum 35 μm metal line has risen, obviously, 3.5-fold compared with the 10 μm Pt line and the same applies to the respective product $m \times c_S$. However, because the Pt substance did not change, the thermal conductivity remained constant. According to **Table 3** degradation for the device #4 (Pt(50 nm) 35 μm wide inert electrode), electrical probing of the Pt(50 nm) line with a width of 35 μm, the surrounding cells' degradation is less severe than that of the Pt(50 nm) arrays' 10 μm Pt lines but slightly worse than that of the Pt(50 nm)/Cu(100 nm) 10 μm wide lines. At this point, it is instructive to revisit **Figure 4** where it is shown that excessive heating leads to damaged (burned) cells. Both cells were operated under identical RESET-SET conditions featuring a Cu filament of the same resistance. The cell at the cross-section of 6 μm Cu and 6 μm Pt line shows severe damage just after three switching cycles. It is seen that the section of Cu electrode covering the cell has completely melted away. The cell at the cross-section of 10 μm Cu and 10 μm Pt line becomes inoperable after six of identical RESET-SET heating cycles and shows only a local melting of Cu along the ridge of the Pt line. The observation provided in **Figure 4** confirms our thermal analysis: the wider Cu line provides not only more thermal mass to accommodate the temperature and therefore the maximum temperature decreases by 40%, but also larger heat loss to the heated cell because of larger heat transfer along the Cu electrode line. Therefore, more heating cycles are needed to cause localized Cu melting on a 10 μm wide electrode lines than on 6 μm wide lines.

Additionally, further support for our thermal analysis comes from earlier research [19], where it has been discovered that a ReRAM using Rh(50 nm)/Cr(30 nm) electrodes in place of Pt(50 nm)/Ti(30 nm) electrodes has nearly identical electric properties to a Cu/TaO_x/Pt/Ti memory cell. But, in contrast, for Pt/Ti devices, a cell set at I_{cc} = 10 μA with M_x of 15 grows to M_x = 26 for Rh/Cr devices. Because of higher thermal conductivity, the device Rh/Cr needs more of the identical SWC to reach the same T_c at which the device becomes inoperable. This is in line with our explanation of the thermal effects. In this particular case, we recall that the only purpose of Cr and Ti layers of 30 nm was to act as suitable adhesion layers. But the two inert electrodes have different thermal conductivities of 150 W/mK and 94 W/mK for Rh/Cr, compared to 69 W/mK and 18 W/mK for the Pt/Ti bilayer combination. According to our thermal investigation, the $m \times c_s$ of the Rh and Pt electrodes are comparable, with the Rh electrode's product and being only marginally greater than the corresponding product for the Pt metal line (see **Table 1**). The product is about the same because Pt has 1.75 times higher mass density but 1.80 times lower specific heat capacity. As, however, Rh's thermal conductivity is 2 times greater than Pt's, we anticipate that using Rh as the electrode will result in cooler surrounding cells than using Pt as an inert electrode. The discovered boost of 26/15 = 1.73 is within a reasonable interval of the estimates provided by our thermal investigation set forth in this work. In a way, the case of the narrower (10 μm wide) Rh(50 nm) electrode is the opposite of the case of the wider Pt(35 μm) electrode. The product $m \times c_s$ grew by a factor of 3.5 when the Pt line was wider, but the thermal conductivity remained constant when compared to a Pt line of 10 μm width. For the Rh 10 μm electrode, the product $m \times c_s$ is the same, but the thermal conductivity value has doubled.

In addition, different adhesive layers Ti and Cr have been added to two identical devices (Cu/TaO_x/Ru) in **Figure 2** (compare inert electrodes (e) and (g)). As opposed to Ti, which has a thermal conductivity of 18 W/mK, Cr has a thermal conductivity of 94 W/mK, i.e., about five times higher. One discovers that the type of adhesive layer sensitively affects the electric performance of apparently equivalent devices. We discover that the Ru/Cr device's $V_{res} = -3.8$ V is 0.4 V higher than the Ru/Ti device's (V_{res}(Ru/Ti) = −3.4 V). According to our analysis, this outcome should have been readily anticipated. Compared to the Ru/Ti device, the Ru/Cr device sees a substantially higher rate of heat conduction from the Cu CF when the device is being heated during the RESET procedure. When using the Ru/Ti electrode, the heat produced in the Cu CF is not released quickly and remains in the area near the hot point for a while because of Ru/Ti's low thermal conductivity. As a result, inert electrodes with lower thermal conductivities require less heating SWC than those with higher thermal conductivities because heat accumulates in the nanofilament at a faster rate with lower thermal conductivities, resulting in lower V_{res} voltage. A comparison of V_{res} values of the other devices is also useful. We discover that the V_{res} for the Co/Ti and Pt/Ti devices are, respectively, V_{res}(Pt/Ti) = −0.9 V and V_{res}(Co/Ti) = −1.0 V, i.e., statistically the same, which, once more, matches extremely well with the identical thermal conductivities of the two metals of 69 W/mK for both Pt and Co. Last but not least, there is a significant variance in V_{res} of around 2.5 V between Pt and Co devices and Ru devices, which is perfectly consistent with Ru's significantly higher heat conductivity than that of cobalt.

As a result, the impact of cell-to-cell heat transfer on the electrical performance of the device is shown to be influenced almost in equal parts by three mechanisms: (i) the electrode materials' specific heat capacity, (ii) thermal conductivity, and (iii) the product of volume and mass density of the electrodes, i.e., its total mass. Our thermal guidelines are able to predict quantitatively the amount of improvement or deterioration of the electrical degradation effects with reference to thermal material properties and geometrical shape of the electrodes, once our thermal methodology has been calibrated to a baseline array (here Cu/TaO_x/Pt/Ti device). Finally, we would like to address the issue of maximum SWC for different devices. Cu/TaO_x/Ru and Cu/TaO_x/Pt devices exhibit identical behavior at first, despite having differing V_{form}, V_{set}, and V_{res} threshold voltages, as demonstrated in Ref. [26]. The basic distinction between the two devices is the severely reduced cycling capability of the Ru devices. The Pt device may cycle through up to 100 SWC under optimal switching conditions, compared to the Ru device's mere 13 cycles. It was discovered that the Ru electrode's decreased inertness qualities were responsible for the reduced switching cycling ability. Similar to this, it was demonstrated in [38] that Cu/TaO_x/Ti and Cu/TaO_x/Ta devices had much lower switching capabilities than Cu/TaO_x/Pt devices.

5. Conclusion

We have provided a wealth of experimental data showing how the material properties and geometrical dimensions of the ReRAM electrodes affect the amount of cell-to-cell heat transfer, and, in turn, how the heat transfer affects the several appearances of degradation of the electrical performance of the resistive memory cells. The degree of device degradation has been quantified using a thorough thermal methodology of the heat transfer that takes into account the electrode materials' thermal conductivity, thermal diffusivity, specific heat capacity, and mass density. Although our memory arrays manufactured in a university laboratory match commercial ReRAM architecture only in terms of the vertical dimension, i.e., the thicknesses of the constitutive layers but exceed commercial products significantly in lateral dimensions, they still provide a great starting point for a reliable source of the essential physics and methodology true also for commercial ReRAM because the thermal reliability issues are driven by the same physical mechanisms and the same fundamental material properties. Actually, the heat density dissipated in state-of-the-art memory arrays is far worse than in our memory arrays, as demonstrated in [20], being about two orders of magnitude greater. The reliability problems and ReRAM degradation brought on by heat-cross-talk may thus be lessened with the help of the thermal analysis put forward here, which can act as a reliable road map for a suitable choice of materials or material combinations. There is currently no satisfactory answer to the question of what might be the best options for the inert electrode as far as their thermal properties are concerned. Low conductivity materials result in lower V_{res} value, which is obviously advantageous for the reduction of the heat dissipated in the arrays, and distant devices are mostly unaffected by the reduced thermal cross-talk alongside the metallization lines. However, for low thermal conductivity electrodes, the Joules heat dissipated in a device lingers longer within the electrode materials and cell. In contrast, the rapid heat transport in the case of a high thermal conductivity electrode impacts by dint of the thermal cross-talk even the cells farthest from the heated cell. ReRAM arrays must be optimally designed and operated taking into account how the memory is being used in terms of programming and erasure cycles to minimize adverse thermal cross-talk effects.

Author details

Amrita Chakraborty, Mohammad S. Al-Mamun and Marius K. Orlowski*
Bradley Department of Computer and Electrical Engineering, Virginia Tech, Blacksburg, USA

*Address all correspondence to: m.orlowski@vt.edu

References

[1] An H, Al-Mamun MS, Orlowski MK, Liu L, Yi Y. Robust deep reservoir computing through reliable Memristor with improved heat dissipation capability. IEEE Transactions on Computer-Aided Design of Integrated Circuits and Systems. 2021;**40**(3):574-583. DOI: 10.1109/TCAD.2020.3002539

[2] Liu T, Kang Y, El-Helw S, Potnis T, Orlowski M. Physics of the voltage constant in multilevel switching of conductive bridge resistive memory. Japanese Journal of Applied Physics. 2013;**52**:084202. DOI: doi.org/10.7567/JJAP.52.084202

[3] De S, Higgins TM, Lyons PE, Doherty EM, Nirmalraj PN, Blau WJ, et al. Silver nanowire networks as flexible, transparent, conducting films: Extremely high DC to optical conductivity ratios. ACS Nano. 2009;**3**: 1767-1774

[4] Hu LB, Kim HS, Lee JY, Peumans P, Cui Y. Scalable coating and properties of transparent, flexible, silver nanowire electrodes. ACS Nano. 2010;**4**: 2955-2963

[5] Yang LQ, Zhang T, Zhou H, Price SC, Wiley BJ, You W. Solution-processed flexible polymer solar cells with silver nanowire electrodes. ACS Applied Materials & Interfaces. 2011;**3**: 4075-4084

[6] Chang Y-F, editor. Memristor—An Emerging Device for Post-Moore's Computing and Applications. London, UK: Intechopen; 2021

[7] Azghadi MF, Chen YC, Eshraghian JK, Chen J, Lin CY, Amirsoleimani A, et al. Complementary metal oxide semiconductor and Memristive hardware for neuromorphic computing. Advanced Intelligent Systems. 2020;**2**:1900189

[8] Yu Z, Zhang Q, Li L, Chen Q, Niu X, Liu J, et al. Highly flexible silver nanowire electrodes for shape-memory polymer light-emitting diodes. Advanced Materials. 2011;**23**:664

[9] Zeng XY, Zhang QK, Yu RM, Lu CZ. A new transparent conductor: Silver nanowire film buried at the surface of a transparent polymer. Advanced Materials. 2010;**22**:4484-4488

[10] Sun P, Lu N, Li L, Li Y, Wang H, Lv H, et al. Scientific Reports. 2015;**5**: 13504

[11] Waser R, Dittmann R, Staikov G, Szot K. Redox-based resistive switching memories—Nanoionic mechanisms, prospects, and challenges. Advanced Materials. 2009;**21**(25–26):2632-2663. DOI: 10.1002/adma.200900375

[12] Russo U, Ielmini D, Cagli C, Lacaita AL. IEEE Transactions on Electron Devices. 2009;**56**:193

[13] Ielmini D. IEEE Transactions on Electron Devices. 2011;**58**:4309

[14] Mosso N, Prasmusinto A, Gemma A, Drechsler U, Novotny L, Gotsmann B. Quantized thermal conductance in metallic heterojunctions. Applied Physics Letters. 2019;**114**:123102. DOI: 10.1063/1.5086483

[15] Chang SH, Chae SC, Lee SB, Liu C, Noh TW, Lee JS, et al. Applied Physics Letters. 2008;**92**:183507

[16] Walczyk C, Walczyk D, Schroeder T, Bertraud T, Sowinska M, Lukosius M, et al. IEEE Transactions on Electron Devices. 2011;**58**:3124

[17] Terai M, Saitoh M, Nagumo T, Sakotsubo Y, Yabe Y, Takeda K, et al. Symposia on VLSI Technology. Kyoto, Japan: IEEE Explore. Jun 14-16, 2011

[18] Lee J-S, Lee S, Noh T-W. Resistive switching phenomena. Applied Physics Reviews. 2015;**2**:031303

[19] Al-Mamun M, Orlowski M. Performance degradation due to nonlocal heating effects in resistive ReRAM memory arrays. MRS Advances. 2019;**4**(48):2593-2600

[20] Al-Mamun M, Orlowski M. Performance degradation of Nanofilament switching due to joule heat dissipation. Electronics MDPI. 2020; **9**(1270):2-11

[21] Al-Mamun M, Orlowski M. Electric conductivity of remotely heated Cu nanofilaments in Cu/TaOx/Pt ReRAM cells. Journal of Applied Physics. 2021;**129**:055107. DOI: 10.1063/5.0034754

[22] Al-Mamun M, Orlowski M. Electron tunneling between vibrating atoms in a copper nano-filament. Scientific Reports. 2021;**11**:7413

[23] Kang Y, Liu T, Potnis T, Orlowski MK. ECS Solid State Letters. 2013;**2**(7):Q54-Q57

[24] Liu T, Verma M, Kang Y, Orlowski M. Applied Physics Letters. 2012;**101**:073510

[25] Sun P, Li L, Lu ND, Li YT, Wang M, Xie HW, et al. Physical model of dynamic joule heating effect for reset process in conductive-bridge random access memory. Journal of Computational Electronics. 2014;**13**(2): 432-438. DOI: doi.org/10.1007/s10825-013-0552-x

[26] Chakraborty A, Al-Mamun M, Orlowski M. Inertness, and other properties of thin ruthenium electrodes in ReRAM applications. In: Chang YF, editor. Ruthenium—Materials Properties, Device Characterizations, and Advanced Applications. London, UK: IntechOpen; 2023. DOI: 10.5772/intechopen.110531

[27] Fangohr H, Chernyshenko DS, Franchin M, Fischbacher T, Meier G. Joule heating in nanowires. Physical Review. 2011;**84**:054437

[28] Uenuma M, Ishikawa Y, Uraoka Y. Joule heating effect in nonpolar and bipolar resistive randow access memory. Applied Physics Letters. 2015;**107**: 073503

[29] Sato Y, Kinoshita K, Aoki M, Sugiyama Y. Consideration of switching mechanisms of binary metal oxide resistive junctions using a thermal model. Applied Physics Letters. 2007;**90**: 033503

[30] Al-Mamun M, Chakraborty A, Orlowski M. Analysis of the electrical ReRAM device degradation induced by thermal cross-talk. Advanced Electronic Materials. 2023;**9**:2201081

[31] Linn E, Rosezin R, Kügeler C, Waser R. Complementary resistive switches for passive nanocrossbar memories. Nature Materials. 2010;**9**: 403-406. DOI: 10.1038/nmat2748

[32] Ghosh G, Orlowski M. Write and erase threshold voltage interdependence in resistive switching memory cells. IEEE Transactions on Electron Devices. 2015; **62**:2850-2856. DOI: 10.1109/TED.2015.2452411

[33] Ghosh G, Orlowski MK. Correlation between set and reset voltages in resistive RAM cells. Current Applied

Physics. 2015;**15**:1124-1129. DOI: 10.1016/j.cap.2015.06.015

[34] Fan Y, Al-Mamun M, Conlon B, King S, Orlowski M. Resistive switching comparison between Cu/TaOx/Ru and Cu/TaOx/Pt cells. ECS Transactions. 2017;**75**(32):13-23

[35] Rafols I, Ortin J. Heat conduction in a metallic rod with Newtonian losses. American Journal of Physics. 1992;**60**(9): 846-852

[36] Lienhard JH, Lienhard JH. A Heat Transfer Textbook. 5th ed. Dover Publications; 2019

[37] Stein EM, Shakarchi R. Fourier Analysis. Princeton University Press; 2003. pp. 750-770

[38] Ghosh G, Kang Y, King S, Orlowski M. Role of CMOS back-end metals as active electrodes for resistive switching in ReRAM cells. ECS Journal of Solid State Science and Technology. 2017;**6**(1):N1-N9

Section 3

Applications

Chapter 6

Application of Two General Memristor Models in Chaotic Systems

Xiangliang Xu, Tianxian Zhang, Yongqi Zhao, Wanying Dai, Si Wu and Guodong Li

Abstract

The memristor has attracted more and more attention due to its broad application prospect. The theory of memristors is being gradually improved. The first is to derive a general voltage-controlled discrete memristor (DM) model from another perspective-circuit, which proves that it conforms to the definition of Memristor, and a Simulink model of the discrete memristor is built to implement the simulation. The other one is the conventional magnetron titanium dioxide (i.e., D) memristor model. In order to explore the adaptability of these two types of memristor models, this paper constructs two types of memristor chaotic systems based on them and performs dynamics analysis to verify the applicability of the above memristor models, which lays the foundation for the application of memristor chaotic systems in the field of communication security.

Keywords: discrete memristor, magnetically controlled titanium dioxide memristor, chaotic system, Simulink modeling, dynamics analysis

1. Introduction

The concept of memristor was proposed by Prof. Chua from the University of California, Berkeley in 1971 based on the completeness of circuit theory, which represents the relationship between magnetic flux and electric charge. The memristor is also considered to be the fourth basic circuit element in addition to resistance, capacitance, and inductance [1, 2]. In 2008, Strukov et al. from Hewlett-Packard Laboratories first developed a nano-memristor device when doing TiO_2 cross-array experiments [2], and the experiments confirmed that the nano-memristor has a unique switching mechanism, automatic memory function, continuous input and output characteristics, and the emergence of the memristor opens up a new space for the development of the design and application of electronic circuits. Because of the memory function of the memristor, its potential application value has attracted extensive attention from academia and industry. Memristors will show attractive application prospects in fields such as neural networks [3], electrical engineering [4], and secure communications [5, 6]. In addition, memristor can mimic the associative learning and forgetting properties of the human brain and thus is used in neuromorphic circuits and artificial intelligence [7].

Since the development of the nano-memristor in HP Labs, various models of the memristor have been proposed. For example, the nonlinear drift model of HP memristor in literature [8], and the flux-controlled memristor model focusing on threshold characteristics in literature [9]. These models are based on ion migration, and although they are consistent with the basic properties of the memristor, many practical behavioral properties are still not represented. In particular, the ion mobility model can simulate the nonlinear transmission properties of synapses when the memristor acts as a synapse, but it is unable to explain the phenomena of long-term plasticity (LTP), short-term plasticity (STP), and the memory fading phenomena in biological systems, which is a common shortcoming of the current ion mobility models. This is a common shortcoming of current ion mobility models. Therefore, it is necessary to find a mathematical model that can fit the above phenomena.

It is worth noting that references [10, 11] replace the Chua diode in the Chua oscillator with a segmented linear memristor, and propose several memristor chaotic oscillation circuits based on Chua circuits. Driven by this groundbreaking work, many scholars have devoted themselves to the study of various memristor chaotic systems [12–15]. The chaotic signal generator is the foundation of chaotic secure communication. Generally speaking, the nonlinearity of the circuit is a necessary condition for chaos generation, so it is necessary to make the system have more complex chaotic behavior. As a controllable nonlinear device, memristors, coupled with their small size and low power consumption, are very suitable for application in high-frequency chaotic circuits. High-frequency chaotic signals have broad application prospects in image encryption [16] and chaotic secure communication [17]. Therefore, the use of memristors to construct chaotic systems has received close attention from researchers, and research on the design and dynamic behavior analysis of memristor chaotic systems has become a hot topic [12, 14].

Since chaos theory has been proposed, the research results on the design of chaotic systems, dynamics analysis, chaos-based nonlinear circuits, and their applications have been quite mature. With the continuous deepening of the application aspects, the requirements for the system complexity are getting higher and higher. The memristor, as a kind of nonlinear element with memory function and nanometer size, also the nonlinear part of the chaotic system, not only can greatly improve the chaotic system's signal randomness and complexity but also reduce the physical size of the system. On the one hand, some scholars have tried to enhance the chaotic properties of chaotic systems by incorporating memristor into them [15, 18, 19], such as fractional-order chaotic systems [18, 20, 21], conservative chaotic systems [22, 23], multiscroll chaotic systems [24–26], and fractal chaotic systems [27, 28]. On the other hand, despite the wide range of applications of memristor in nonlinear systems, most of the studies are based on continuous memristor models. However, continuous memristor is difficult to be applied to discrete systems or digital circuits. Therefore, researchers have started to focus on discrete memristors and have achieved some remarkable results in this area. Several researchers have derived discrete memristor models using continuous memristors [29–31]. Literature [30] modeled a discrete memristor using the forward Eulerian difference method and obtained a new two-dimensional hyperchaotic mapping from sampled memristor-capacitor circuits. Literature [31] proposed a discrete memristor model based on the difference theory, proved the three-fingerprint characterization of the model according to the definition of generalized memristor, and applied it to the Hénon mapping. In addition, the discrete HP memristor derived from the continuous HP memristor is the most popular research topic [29, 32, 33]. Literature [29] constructed a multidimensional closed-loop coupling model and then introduced HP discrete memristors into the coupling model. It also constructed a discrete memristor coupling model based on Logistic mapping and sinusoidal mapping, which can produce an arbitrary high-dimensional hyperchaotic system. Literature [32] designed two discrete memristors with cosine band amplitude memristors and established Simulink models of two discrete memristors which verified that they conformed to the definition of memristor [33] so that the dynamics of classical chaotic mapping can be improved. Several scholars have derived general generic models for discrete memristor. Literature [34] proposed a general discrete memristor model and a unified discrete memristor mapping model, also proposed four 2D-discrete memristor hyperchaotic mappings with infinite immovable points, and studied their dynamical behaviors. Literature [35] proposed a generalized voltage-controlled discrete memristor model, improved three typical discrete chaotic mappings, and studied their dynamical behaviors. In summary, memristor is of great significance in enhancing nonlinear systems, especially in the research of combining with chaotic fields, due to its good nonlinear characteristics, and the study and control of memristor chaotic systems is a new issue that needs to be paid attention to in the engineering applications of nonlinear dynamical systems.

Therefore, there is a need to further develop and improve the theory of memristor electrical networks, to explore the nonlinear characteristics of memristor and its role in the complex dynamics and formation mechanism of chaotic systems, to reveal the plasticity of memristor is of great significance in the evolution and function of neural networks [36–38] and neuromorphic circuits [39]. The combination of memristor and chaos can be mainly divided into two aspects: on the one hand, memristor has become one of the hotspots of research due to its inherent nonlinear properties and complex dynamical behaviors [30, 31, 33, 34, 40, 41]. On the other hand, memristor is often used as one of the applications to introduce into other chaotic systems to enhance the chaotic properties due to its memory function and nonlinearity [15, 18, 24, 26, 29, 31, 32, 35].

In the above studies, the memristor model is usually established by defining the memristor's memristance or memductance, or the model is established directly from a specific memristor. However, the ideal memristor model cannot completely simulate the characteristics of the actual device, so the establishment of a general-purpose memristor model is more practical significance, but there has been less exploration of the adaptability of the universal memristor model. In this context, through the relevant theoretical derivation of memristor, this paper proposes two generalized memristor models, gives the specific construction methods and application rules, and applies them to two classical chaotic systems respectively which verifies the correctness and applicability of the proposed models from the experimental results of dynamics analysis.

The rest of the paper is organized as follows. Section 2 mainly proposes two generalized memristor models, namely, the generalized voltage-controlled discrete memristor model and the generalized magnetic-controlled titanium dioxide memristor model, constructs the corresponding memristor chaotic system, analyzes the dynamics of the system, and verifies the correctness and applicability of the proposed memristor models. Section 3 summarizes and discusses the whole paper.

2. Design and application of generalized memristor modeling

In this section, two types of generalized memristor mathematical models are proposed, namely, the generalized voltage-controlled discrete memristor model and the generalized magnetic-controlled titanium dioxide memristor model. By studying the mechanism of action and nonlinear properties inside of the memristor, the above two memristor models are applied to two types of classical chaotic systems: discrete Hénon chaotic mapping and Lorenz chaotic systems, respectively, to verify the universality of the proposed generalized memristor mathematical model.

2.1 Generalized voltage-controlled discrete memristor model and its application to chaotic mapping

2.1.1 Generalized voltage-controlled discrete memristor model

In general, the ideal mathematical model of a magnetic-controlled memristor is defined as:

$$q = \hat{q}(\varphi), \tag{1}$$

or

$$\begin{cases} i(t) = W(\varphi) \cdot v(t), \\ d\varphi(t)/dt = v(t), \end{cases} \tag{2}$$

where φ and q denote the input magnetic flux and the charge flowing through the device, respectively, v is the input voltage, and i is the output current. $W(\varphi)$ is the memductance, which is a nonlinear function of the magnetic flux $\varphi(t)$.

Chua [42] gave the steps of an ideal voltage-controlled memristor through the nonlinear intrinsic relationship between the device terminal voltage v and the terminal current i, moreover, Bao et al. [30] constructed a model of a magnetic-controlled memristor with nonlinear memductance, which is defined as:

$$\begin{cases} i(t) = \cos\varphi(t) \cdot v(t), \\ d\varphi(t)/dt = v(t). \end{cases} \tag{3}$$

Using the above model of the memristor, a sampling memristor-capacitor circuit with a sample interval switch is further proposed to connect the memristor to the capacitor, which is shown in **Figure 1**.

The charge-controlled discrete memristor is derived by introducing the forward difference method into the charge-controlled continuous memristor [43], and similarly, the voltage-controlled discrete memristor can be derived using the difference method [44].

Consider an ideal memristor as shown in Eq. (2). Denote v_n, φ_n and i_n as the capacitor voltage value, the memristor magnetic flux, and the current value at the time n, respectively. In the sampling memristor-capacitor circuit of **Figure 1**, the intrinsic relationship between the memristor and the capacitor can be expressed by the Euler method as follows:

$$\begin{cases} i_n = W(\varphi_n) \cdot v_n, \\ \varphi_{n+1} - \varphi_n = \Delta T \cdot v_n, \end{cases} \tag{4}$$

and

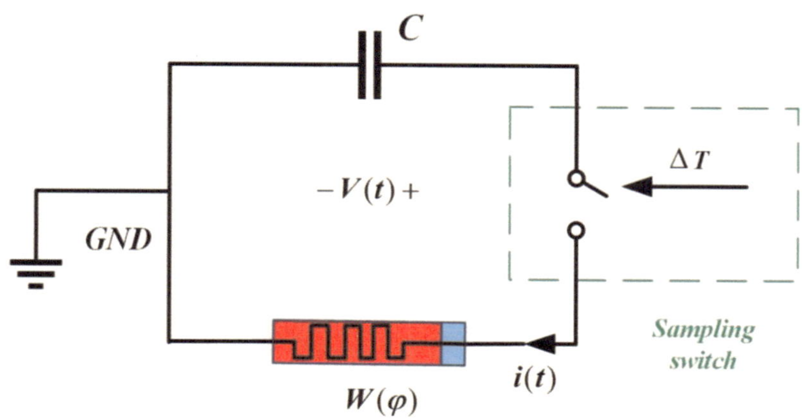

Figure 1.
The sampling memristor-capacitor circuit.

$$C(v_{n+1} - v_n) = -\Delta T \cdot i_n, \tag{5}$$

where C is the capacitance.

This can be obtained by applying Kirchhoff's current law to the circuit in **Figure 1**.

$$\begin{cases} v_{n+1} = v_n - k \cdot W(\varphi_n) \cdot v_n, \\ \varphi_{n+1} = \varphi_n + \varepsilon v_n, \end{cases} \tag{6}$$

where $k = \Delta T / C, \varepsilon = \Delta T$ are two positive parameters, and the relation between v_n and $W(\varphi_n)$ can be obtained from the above equation:

$$\begin{aligned} W(\varphi_n) &= W(\varphi_{n-1} + \varepsilon \cdot v_{n-1}) \\ &= W\left(\varphi_0 + \varepsilon \cdot \sum_{j=0}^{n-1} v_j\right) \\ &= W\left(q_0 + \varepsilon \cdot \sum_{j=0}^{n-1} v_j\right), \end{aligned} \tag{7}$$

where q_0 can be regarded as the initial charge, so that the generalized voltage-controlled discrete memristor model can be obtained as:

$$i_n = v_n \cdot W\left(q_0 + e \cdot \sum_{j=0}^{n-1} v_j\right), \tag{8}$$

where e is a parameter related to the sampling interval, which can also be viewed as a parameter of the discrete memristor.

In the generalized voltage-controlled discrete memristor model, a new discrete memristor model can be constructed by defining memductance, which can satisfy the essential characteristics of the memristor while satisfying the diversity of the memristor. It is worth noting that the steps involved in constructing a new discrete memristor based on the generalized voltage-controlled discrete memristor model are as follows:

Step 1: Select a suitable continuous function as the memductance $W(\varphi)$, to obtain:

$$i_n = v_n \cdot W\left(q_0 + e \cdot \sum_{j=0}^{n-1} v_j\right), \tag{9}$$

Step 2: Apply a sine voltage signal with different frequencies to the above equation, and the volt-ampere characteristic curve can be obtained.

Step 3: The volt-ampere characteristic curve is used to determine whether the constructed discrete memristor satisfies the three basic characteristics of a memristor [45]:

1. Have pinched hysteresis loops at the origin;

2. Frequency dependence of hysteresis loop lobe areas, hysteresis loop lobe areas decrease monotonically with increasing frequency of the periodic excitation signal;

3. When the frequency is infinite, the hysteresis loops contract to a single-valued function.

To verify the applicability of the proposed generalized voltage-controlled discrete memristor model, in this paper, the memductance is defined as a common primitive function, as shown in **Table 1**.

A sine voltage signal $v_n = A_0 \sin(\omega n)$ is used as the input to the discrete memristor model. The volt-ampere characteristic curves at different frequencies are shown in **Table 2**, where it can be observed that the trajectory in the $v - i$ plane is a pinched hysteresis loop at the origin, and the hysteresis loop lobe areas decrease monotonically with increasing frequency of the sinusoidal voltage signal. When the frequency is infinite, the hysteresis loops shrink as a single-valued function. Therefore, the constructed discrete memristor models all have the three fingerprints characteristics of the memristor [31].

Based on the mathematical model of a generalized voltage-controlled discrete memristor, its Simulink model is constructed as shown in **Figure 2**. The model consists of a unit delay block, a summation block, a memory block, an addition block, a gain block, a product block, a constant block, and a $W(\varphi)$ function block. The summing block and the memory block form an accumulator. The $W(\varphi)$ function block represents the function that defines the memductance. Different discrete memristor Simulink models can be created by setting the $W(\varphi)$ function. XY Graph represents the relationship between the input voltage v_n and the memristor current i_n.

Function categories	$W(\varphi)$
Trigonometric functions	$W(\varphi) = \sin\varphi,\ W(\varphi) = \cos\varphi$
Inverse trigonometric function	$W(\varphi) = \arccos\varphi$
Power function	$W(\varphi) = \varphi^{\pi}$
Exponential function	$W(\varphi) = e^{\varphi}$
Logarithmic function	$W(\varphi) = \log\varphi$

Table 1.
Defining functions of memductance in discrete memristors.

$W(\varphi) = \sin\varphi$	$W(\varphi) = \cos\varphi$	$W(\varphi) = \arccos\varphi$
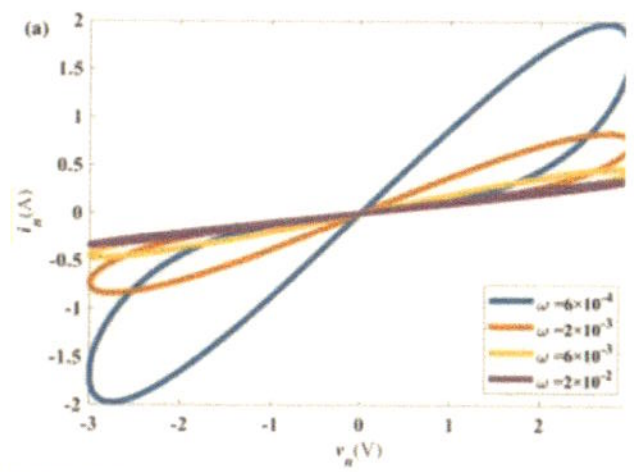	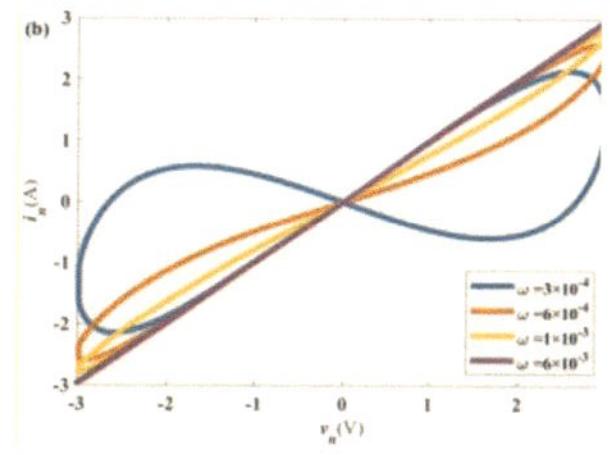	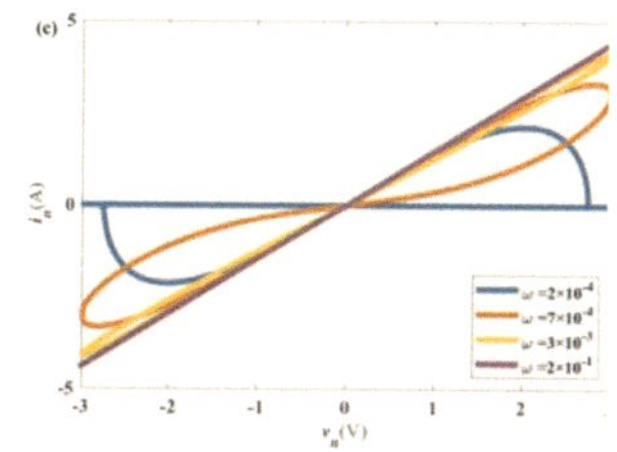
$W(\varphi) = \varphi^{\pi}$	$W(\varphi) = e^{\varphi}$	$W(\varphi) = \log\varphi$
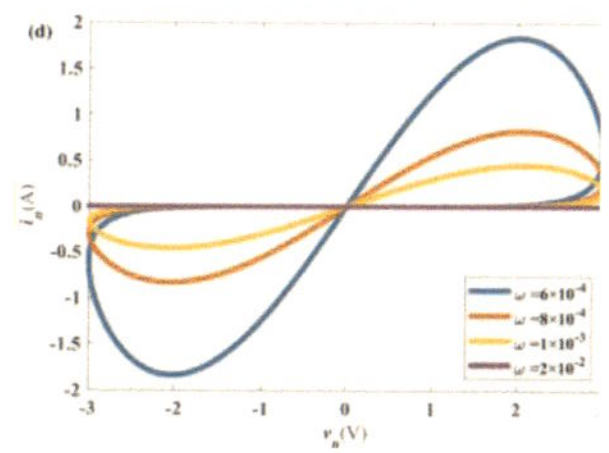	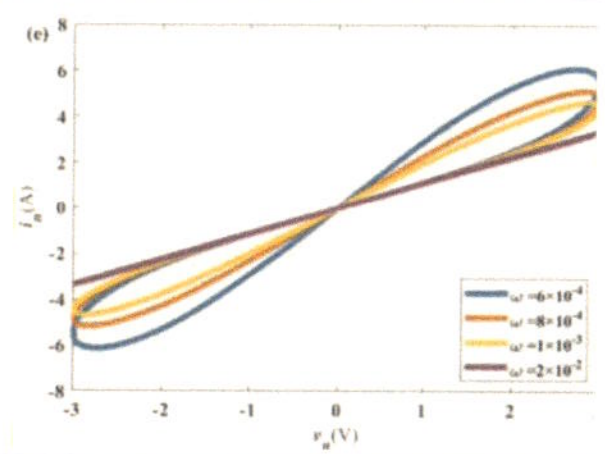	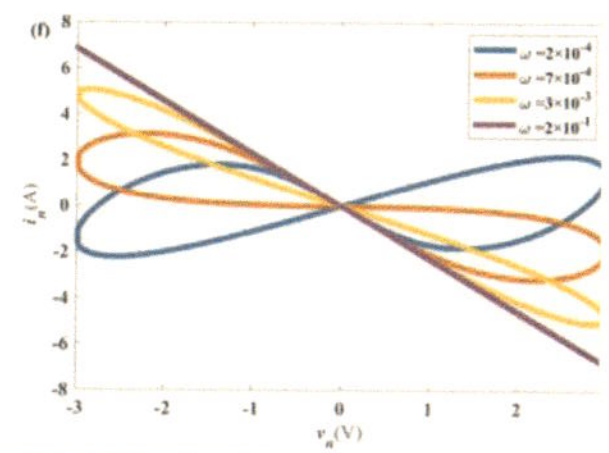

Table 2.
Hysteresis loops of discrete memristors.

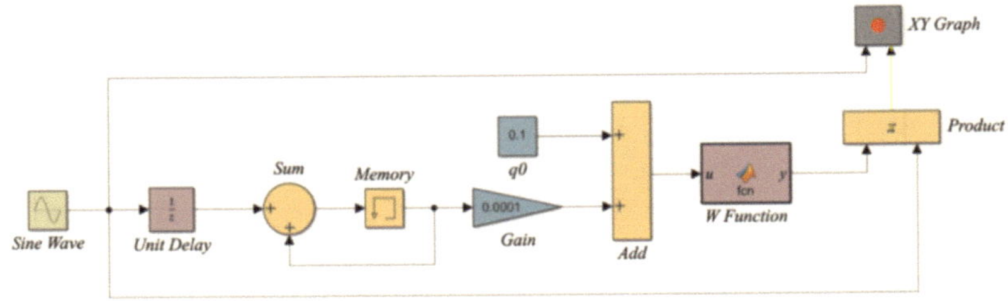

Figure 2.
Simulink model of the generic voltage-controlled discrete memristor model.

To verify the correctness of the Simulink model, the similarity between the Simulink model of the generalized voltage-controlled discrete memristor and the experimental data from the numerical simulation has been measured. The v_n and i_n obtained from the Simulink model will be recorded by using the To Workspace module. Due to the difference in sampling time between the Simulink model and the numerical simulation, the Simulink model sine module is used in the numerical simulation as the input sine voltage signal v_n . The amplitude of the sine voltage is $A = 3$ and the several frequency are set as
$\omega = 2 \times 10^{-4}, 5 \times 10^{-4}, 3 \times 10^{-3}, 2 \times 10^{-2}$, respectively. Here, the $W(\varphi)$ function is set as a logarithmic function. The relationship between the Simulink model and the numerical simulation of the volt-ampere characteristic curve is shown in **Figure 3**. Each volt-ampere characteristic curve corresponds to 400,000 points. The results show that the Simulink model is consistent with the change of hysteresis loop trend of the numerical simulation experiments, thus verifying the correctness of the generalized voltage-controlled discrete memristor model.

2.1.2 Hénon mapping model based on discrete memristor

The Hénon mapping [46] is a common two-dimensional discrete chaotic mapping with the system equation:

$$\begin{cases} x_n = 1 + y_{n-1} - a x_{n-1}^2, \\ y_n = bx, \end{cases} \tag{10}$$

where a and b are the control parameters of the system.

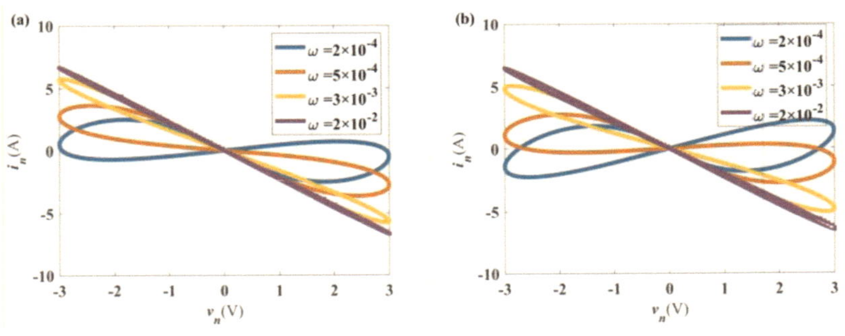

Figure 3.
The pinched hysteresis loops obtained from (a) the Simulink model; (b) the numerical simulation.

For introducing a discrete memristor into the Hénon mapping, there are two choices of the discrete memristor model's inputs, x_n or y_n. Here, x_n is chosen as the input to the voltage-controlled discrete memristor, and the y_{n-1} variable in x_n is chosen to multiply the output of the memristor, let $W(\varphi) = \cos\varphi$. Thus, the Hénon mapping model based on the discrete memristor is as follows:

$$\begin{cases} x_n = \begin{cases} 1 - ax_{n-1}^2 + y_{n-1}x_{n-1}W\left(q_0 + e\sum_{i=1}^{n-1} x_i\right), & 2 \le n \le 100 \\ 1 - ax_{n-1}^2 + y_{n-1}x_{n-1}W\left(q_0 + e\sum_{i=n-99}^{n-1} x_i\right), & n \ge 101 \end{cases} \\ y_n = bx_{n-1}, \\ W(\varphi) = \cos x. \end{cases} \tag{11}$$

Where q_0 and e are respectively the internal initial values and parameters of the discrete memristor. Inspired by the literature [31], the "short-term memory" method is used to limit the memory of the discrete memristor, i.e., for the first 100 items of the sequence, the memristor remembers all the items, and after the first 100 items of the sequence, the memristor only remembers the first 99 items.

2.1.3 Dynamics analysis of Hénon mapping based on discrete memristor

The dynamical behavior of DM-based Hénon mappings, including attractor trajectories, Lyapunov exponential spectrum, bifurcation maps, attractor basins, and sequence complexity, are studied and analyzed to compare and contrast the specific effects of voltage-controlled discrete memristors on Hénon chaotic mappings.

The chaotic attractor is shown in **Figure 4**, setting the initial value of the system Eq. (11) to $(x_0, y_0) = (0.1, 0.1)$ and the internal parameter $q_0 = 0.1, e = 0.0001$ of the memristor model (the same as below), unlike the original chaotic mapping, the trajectory of the attractor is an arc. **Figure 4** shows the attractors of DM-based Hénon chaotic mapping for different parameters. In **Figure 4(a)**, when the fixed parameter $b = 0.3$, the gap of the arcs disappears and the length of the attractor arcs increases as the parameter a increases. When $a = 1.6$, the attractor arc length is the longest. In addition, in **Figure 4(b)**, the folding degree of the attractor decreases with the increase of the parameter b when the fixed parameter $a = 1.5$. It can be seen that when

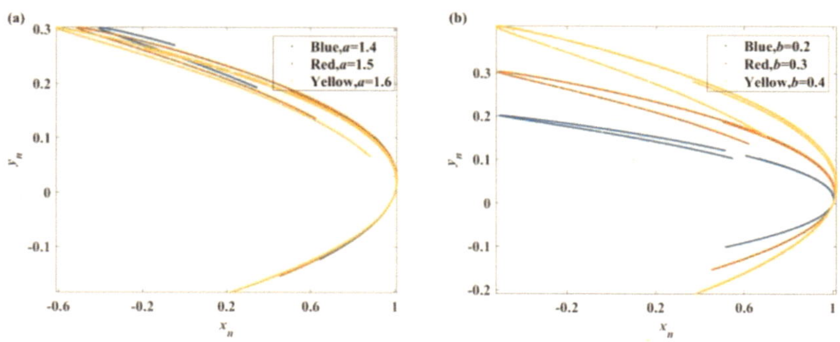

Figure 4.
Chaotic attractor maps for DM-based Hénon mapping. (a) $b = 0.3$. (b) $a = 1.5$.

$b = 0.2$, there is a large gap between the arcs, and the attractor consists of three arcs, and there are two arcs in all the upper part of the attractor, and the two arcs intersect at a point, and the distance between these two arcs increases as the parameter b increases. It can be summarized that the parameter a controls the length of the arcs and the gap between the arcs, while the parameter b controls the number of arcs, the gap between the arcs, the distance between the arcs, and the folding state of the attractor.

The Lyapunov exponential spectrum and its corresponding bifurcation under different parameters control for the Hénon mapping and the DM-based Hénon mapping are shown in **Figures 5** and **6**, respectively. **Figure 5** shows the Lyapunov exponential spectrum and bifurcation maps of the corresponding systems under the control of the parameter a. The Hénon mapping has the system in a chaotic state at $a \in [1.15, 1.607]$, while the DM-based Hénon mapping has the system in a chaotic state at $a \in [1.354, 1.953]$. The chaotic region of the DM-based Hénon mapping is wider than that of the original Hénon mapping, a situation that is more evident in the parameter b. **Figure 6** shows the Lyapunov exponential spectrum and the bifurcation diagram for the corresponding system under the control of the parameter b. It can be observed that the LE value of the DM-based Hénon mapping is almost always greater than zero within $b \in [0, 0.9]$, and the Hénon mapping has a LE value greater than zero only within $b \in [0, 0.313]$. Therefore, it can be concluded that the voltage-controlled discrete memristor effectively expands the chaotic parameter space of the Hénon mapping.

To further investigate the effect of the initial state values, an attraction basin consisting of the $x_0 - y_0$ plane is shown in **Figure 7**, where the white region is the

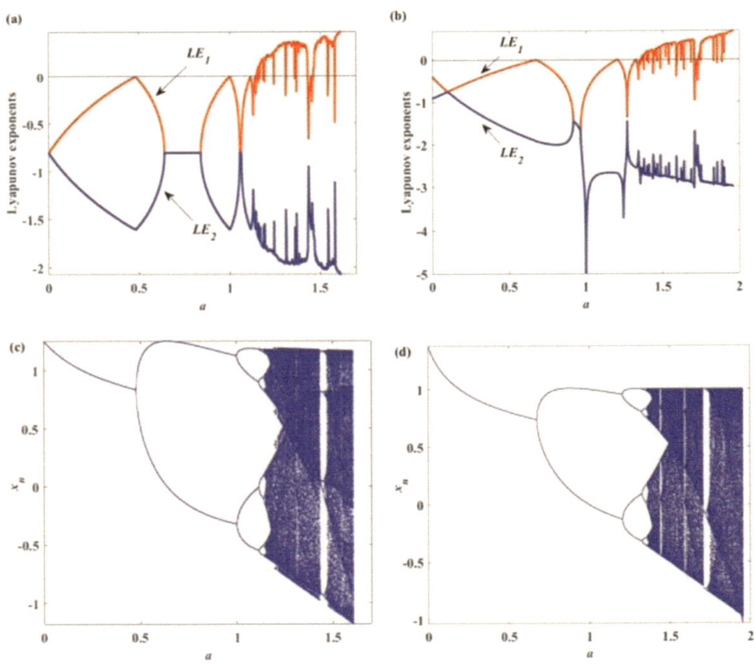

Figure 5.
Lyapunov exponential spectrum and bifurcation diagrams of Hénon mapping and DM-based Hénon mapping $x_0 = 0.0001, y_0 = 0.0001$. (a) Lyapunov exponential spectrum of Hénon mapping $a \in [0, 1.7], b = 0.2$. (b) Lyapunov exponential spectrum of DM-based Hénon mapping $a \in [0, 2], b = 0.2$. (c) Bifurcation diagram of Hénon mapping $a \in [0, 1.7], b = 0.2$. (d) Bifurcation diagram of DM-based Hénon mapping $a \in [0, 2], b = 0.2$.

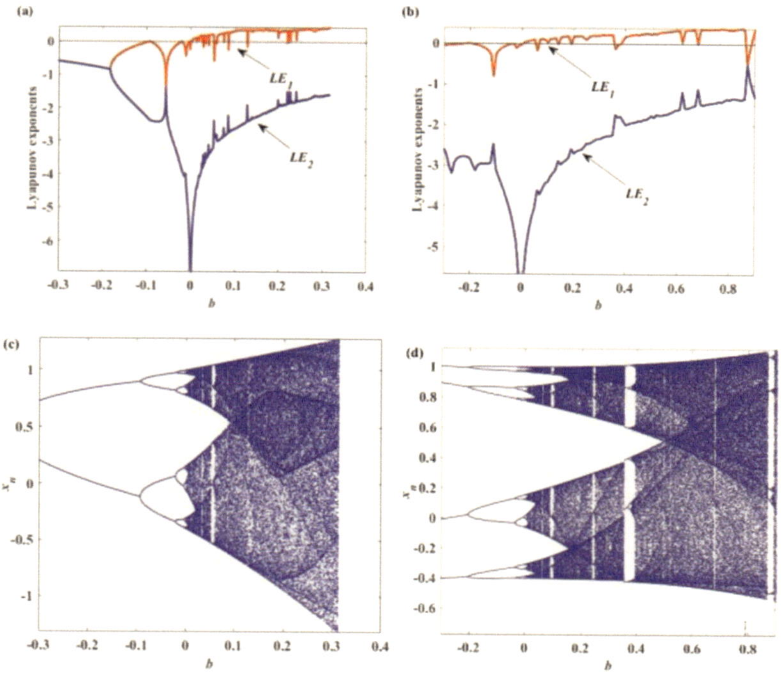

Figure 6.
Lyapunov exponential spectrum and bifurcation diagrams of Hénon mapping and DM-based Hénon mapping $x_0 = 0.0001, y_0 = 0.0001$. *(a) Lyapunov exponential spectrum of Hénon mapping* $b \in [-0.3, 0.4], a = 1.4$. *(b) Lyapunov exponential spectrum of DM-based Hénon mapping* $b \in [-0.3, 0.9], a = 1.4$. *(c) Bifurcation diagram of Hénon mapping* $b \in [-0.3, 0.4], a = 1.4$. *(d) Bifurcation diagram of DM-based Hénon mapping* $b \in [-0.3, 0.9], a = 1.4$.

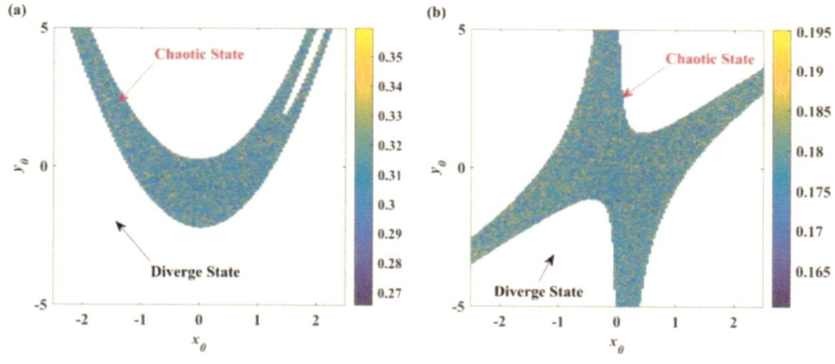

Figure 7.
Attraction basin in the $x_0 - y_0$ *plane. (a) Hénon mapping* $a = 1.4$, $b = 0.2$. *(b) DM-based Hénon mapping* $a = 1.4, b = 0.2$.

divergence region, i.e., the sequence generated by the system is divergent at the corresponding initial value of the point, and the non-white region is the chaotic region. It is worth noting that the voltage-controlled L effectively extends the valid region of the initial values x_0 and y_0 of the original system, and the valid region of the initial values of the DM-based Hénon mapping is larger than that of the original Hénon mapping, which suggests that the discrete memristor can change and extend the range of the initial values of the chaotic region of the Hénon mapping.

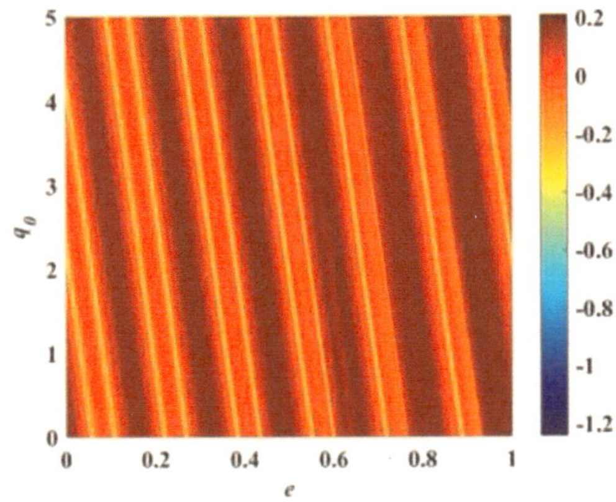

Figure 8.
Chaotic map of $e - q_0$ plane based on Lyapunov exponent $e \in [0, 1], q_0 \in [0, 5]$, $a = 1.4, b = 0.2, x_0 = 0.1, y_0 = 0.1$.

To investigate the effect of the internal parameters of the memristor on the system's degree of chaos, the parameters and initial values of the system are fixed, and the internal parameters of the memristor, $e \in [0, 1], q_0 \in [0, 5]$, are chosen to be 600×600 samples. **Figure 8** shows the Lyapunov exponential chaotic map based on the Hénon mapping of the DM. It can be observed that with the changes of e and q_0, the chaotic region (dark red) and the non-chaotic region (yellow) are distributed in parallel lines, and the system's degree of chaos will change with the selection of the internal parameters of the memristor, which indicates that the nonlinear characteristics of the discrete memristor will have a more obvious effect on the degree of chaos of the Hénon mapping.

Figure 9 shows the Permutation Entropy (PE) of the system under two-parameter control, i.e., from the point of view of the entropy value of the system sequence to determine the system's degree of chaos, the red and yellow regions indicate that the system's PE value is larger, i.e., the system parameters can obtain a high-complexity sequence by selecting the corresponding region, and at this time, the system's degree of chaos is larger. It can be noted that in **Figure 9(a)**, the Hénon mapping is only in the interval of $a \in [0, 2], b \in [0, 1]$ when the PE complexity of the system is greater than

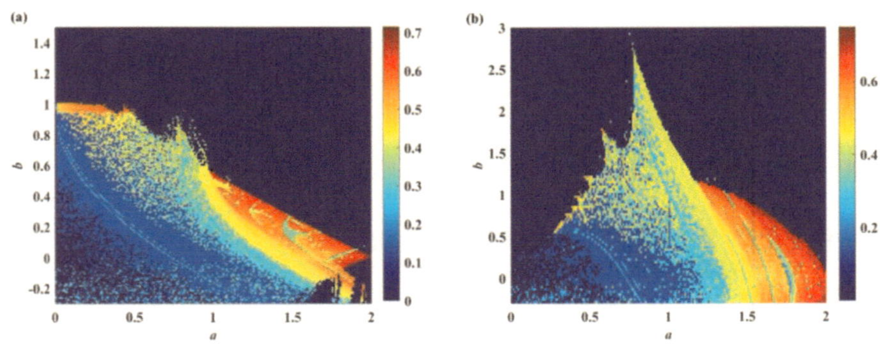

Figure 9.
Permutation entropy with parameters a, b and $x_0 = 0.0001, y_0 = 0.0001$. (a) Hénon map with $a \in [0, 2], b \in [-0.3, 1.5]$. (b) DM-based Hénon map with $a \in [0, 2], b \in [-0.3, 3]$.

	Chaos range of a	Chaos range of b	Initial value chaos ratio	High PE radio
Hénon mapping	0.457	0.313	20.14%	22.43%
DM-based Hénon mapping	0.599	0.9	30.36%	36.08%

Table 3.
Comparison of DM-based Hénon mapping and Hénon mapping key metrics.

0. The PE value in the dark blue region is equal to 0 when the system is in the divergence state. In **Figure 9(b)**, on the contrary, the complexity region of the system Eq. (11) has increased significantly compared to system Eq. (10), which is consistent with the results of the multiple dynamics analysis methods described above. The results show that based on the voltage-controlled discrete memristor proposed in this paper, the parameter range and initial value range of the chaotic mapping as well as the robustness of the strengthened chaotic system can be effectively enhanced, which is more applicable to the field of communication encryption.

A comparison of the main metrics of the DM-based Hénon mapping discussed in this section with the classical Hénon mapping is given in **Table 3**, highlighting the fact that the DM-based Hénon mapping exhibits significant advantages in all the main metrics, with a larger chaotic range.

2.2 Generalized HP memristor model and its application in chaotic systems

2.2.1 Generalized HP memristor model

Firstly, reviewing the structure and definition of the HP memristor model, Strukov et al. found that a layer of titanium dioxide (TiO_2) film was sandwiched between two platinum electrodes, and that the TiO_2 film is divided into two parts, one being a pure TiO_2 thin film layer, which has high resistivity, and the other being a TiO_{2-X} thin film layer. Due to the "doping" of pure TiO_2, some of the oxygen atoms are missing, so the TiO_{2-X} film layer is positively charged and has a lower resistance to the passage of current. When current flows through the device in a counterclockwise direction, the boundary between the two materials shifts, thereby increasing the proportion of TiO_{2-X} film layers, at which point the overall resistance decreases. When the current is reversed, the boundary moves in the opposite direction and the proportion of pure TiO_2 layers increases, increasing the overall resistance. When the current stops, the boundary between the two film layers stops moving and maintains its final resistance value. In other words, the memristor "remembers" the current flowing through it. The physical model of a memristor [2] is shown in **Figure 10**. The corresponding mathematical model is shown as:

$$\begin{aligned} M(t) &= R_{\text{ON}}\frac{w(t)}{D} + R_{\text{OFF}}\left(1-\frac{w(t)}{D}\right), \\ &= R_{\text{OFF}} + (R_{\text{ON}} - R_{\text{OFF}})\frac{w(0)}{D}, \end{aligned} \tag{12}$$

where $M(t)$ is the memristor value, R_{OFF} and R_{ON} are the two limiting values of the memristor resistance, $w(t)$ is the time-dependent thickness of the TiO_{2-X} film layer,

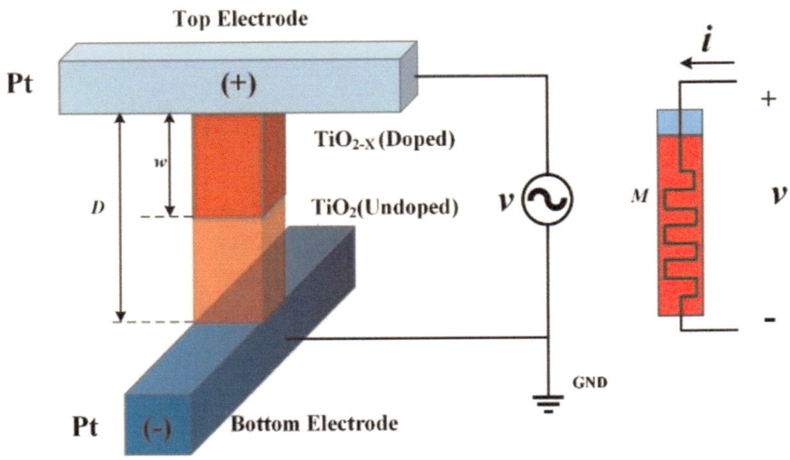

Figure 10.
Schematic of HP memristor.

and D is the thickness of the $Pt - TiO_{2-X} - TiO_2 - Pt$ device film. The R_{OFF} value corresponds to the resistance of the device at $w(t) = 0$, and the R_{ON} value corresponds to the device resistance at $w(t) = D$. At this time, the linear drift model of the device can be represented as:

$$\frac{dw(t)}{dt} = \mu_V \frac{R_{ON}}{D} i(t), \tag{13}$$

Where μ_V is the average mobility of oxygen vacancy.

To derive the complete mathematical model of the HP memristor, it is necessary to consider the interdependence between memristor, charge, and magnetic flux as well as the boundary and initial conditions of the operation. Usually, the initial value of the memristor is not 0, i.e., $w(t)|_{t=0} \neq 0$, and $w(0)$ is set as the initial thickness of $w(t)$, when the initial state value of the memristor is:

$$M(0) = R_{OFF} + (R_{ON} - R_{OFF})\frac{w(0)}{D}, \tag{14}$$

which is obtained by integrating Eq. (13):

$$w(t) = \mu_V \frac{R_{ON}}{D} q(t) + w(0), \tag{15}$$

thus Eq. (12) can be rewritten as:

$$M(t) = M(0) + \frac{(R_{ON} - R_{OFF})\mu_V R_{ON}}{D^2} q(t), \tag{16}$$

i.e.:

$$M(t) = M(0) + k \cdot q(t), \tag{17}$$

where $k = \frac{(R_{ON} - R_{OFF})\mu_V R_{ON}}{D^2}$, and is a constant.

Consider that there exists a boundary condition $w(t) \in [0, D]$ for the memristor, i.e., $R_{ON} \le M(t) \le R_{OFF}$. From Eq. (17), the linear relationship between $q(t)$ and $M(t)$ is given by:

$$\frac{R_{OFF} - M(0)}{k} \le q(t) \le \frac{R_{ON} - M(0)}{k}. \tag{18}$$

let

$$\frac{R_{OFF} - M(0)}{k} = c_1, \quad \frac{R_{ON} - M(0)}{k} = c_2. \tag{19}$$

Since $R_{ON} \ll R_{OFF}$, thus $k < 0$, Eq. (17) is a monotonically decreasing function, the charge control model of the HP memristor is a continuous function with the following mathematical expression:

$$M(t) = \begin{cases} R_{OFF}, & q(t) < c_1 \\ M(0) + kq(t), & c_1 \le q(t) < c_2 \\ R_{ON}, & q(t) \ge c_2. \end{cases} \tag{20}$$

Assuming that an input signal, i.e., $q(0) = 0$, is applied to the device at the moment $t = 0$, the relationship between charge and magnetic flux can be expressed according to the definition of magnetic flux in a memristor:

$$\varphi(t) = \int_0^t M(t) i(t) dt = \int_{q(0)}^{q(t)} M(t) dq(t) \tag{21}$$

then

$$\varphi(t) = \begin{cases} \int_{q(0)}^{c_1} [M(0) + kq(t)] dq(t) + \int_{c_1}^{q(t)} R_{OFF} dq(t) = R_{OFF} q(t) + c_3, & q(t) < c_1 \\ \int_{q(0)}^{q(t)} [M(0) + kq(t)] dq(t) = \frac{k}{2} q^2(t) + M(0) q(t), & c_1 \le q(t) < c_2 \\ \int_{q(0)}^{c_2} [M(0) + kq(t)] dq(t) + \int_{c_2}^{q(t)} R_{ON} dq(t) = R_{ON} q(t) + c_4, & q(t) \ge c_2 \end{cases} \tag{22}$$

where

$$\begin{cases} c_3 = \frac{k}{2} {c_1}^2 + [M(0) - R_{OFF}] c_1 = -\frac{[R_{OFF} - M(0)]^2}{2k}, \\ c_4 = \frac{k}{2} {c_2}^2 + [M(0) - R_{ON}] c_2 = -\frac{[R_{ON} - M(0)]^2}{2k}. \end{cases} \tag{23}$$

Therefore, from Eq. (22), the charge as a function of magnetic flux is given by:

$$q(t) = \begin{cases} \frac{\varphi(t) - c_3}{R_{OFF}}, & \varphi(t) < c_5 \\ \frac{\sqrt{2k\varphi(t) + M^2(0)} - M(0)}{k}, & c_5 \le \varphi(t) < c_6 \\ \frac{\varphi(t) - c_4}{R_{ON}}, & \varphi(t) \ge c_6 \end{cases} \tag{24}$$

where

$$\begin{cases} c_5 = c_1 R_{\mathrm{OFF}} + c_3 = \dfrac{R_{\mathrm{OFF}}^2 - M^2(0)}{2k}, \\ c_6 = c_2 R_{\mathrm{ON}} + c_4 = \dfrac{R_{\mathrm{ON}}^2 - M^2(0)}{2k}. \end{cases} \tag{25}$$

Bringing Eq. (24) into Eq. (20) yields a generalized magnetic control model for HP memristors:

$$M(t) = \begin{cases} R_{\mathrm{OFF}}, & \varphi(t) < c_5 \\ \sqrt{2k\varphi(t) + M^2(0)}, & c_5 \leq \varphi(t) < c_6 \\ R_{\mathrm{ON}}, & \varphi(t) \geq c_6. \end{cases} \tag{26}$$

2.2.2 Modeling of Lorenz chaotic systems based on HP memristor

It is well known that the classical Lorenz chaotic system is a continuous nonlinear system with state equations:

$$\begin{cases} \dfrac{dx}{dt} = \sigma(y - x), \\ \dfrac{dy}{dt} = \rho x - y - xz, \\ \dfrac{dz}{dt} = xy - \gamma z, \end{cases} \tag{27}$$

where σ, ρ, γ are the system parameters, and when $\sigma = 10, \rho = 28, \gamma = \frac{8}{3}$ (the same as below), the system is in a chaotic state.

Without changing the structure of the original system equation of state, the input flux of the HP memristor is used to replace "some single variable" of the Lorenz system equation, i.e., the relationship between charge and magnetic flux in Eq. (24) is utilized to generate complex chaotic signals. It is worth noting that the introduction of a generalized HP memristor model into common nonlinear chaotic systems (which include high-dimensional nonlinear chaotic systems of arbitrary dimensions) requires the following two rules to be followed simultaneously:

Rule 1: Pick any single variable in the system that has a negative coefficient;

Rule 2: In the other two nonlinear differential equations, one differential equation has one and only one selected single variable, and the other differential equation has one and only one cross-product term containing the selected single variable.

Based on the above approach, a Lorenz chaotic system based on HP memristor, i.e., Lorenz memristor chaotic system, is proposed in this paper, and the state equations of the constructed new system are as follow:

$$\begin{cases} \dfrac{dx}{dt} = \sigma(y - x), \\ \dfrac{dy}{dt} = \rho x - f(|y|) - xz, \\ \dfrac{dz}{dt} = xy - \gamma z, \end{cases} \tag{28}$$

where $f(\cdot)$ is the nonlinear term of the system that satisfies the relationship between magnetic flux and charge with the expression:

$$f(y) = \begin{cases} \dfrac{y - c_3}{R_{\text{OFF}}}, & y < c_5 \\ \dfrac{\sqrt{2ky + M^2(0)} - M(0)}{k}, & c_5 \leq y < c_6 \\ \dfrac{y - c_4}{R_{\text{ON}}}, & y \geq c_6 \end{cases} \tag{29}$$

where the y variable in the second differential equation is the input magnetic flux φ of the memristor, and the expressions for $c_i (i = 3, 4, 5, 6)$ are in Section 2.2.1 and will not be repeated here.

2.2.3 Dynamics analysis of Lorenz chaotic system based on HP memristor

The dynamical behavior of the Lorenz memristor chaotic system is studied and analyzed, including the attractor phase diagram, largest Lyapunov exponents (LLE), bifurcation diagram, sequence complexity, etc., to compare and analyze the specific effect of HP memristor on Lorenz chaotic system.

$R_{\text{OFF}}, R_{\text{ON}}$ are the two limit values of the memristor, set $R_{\text{OFF}} = 20\text{k}\Omega$, $R_{\text{ON}} = 100\Omega$, $M(0) = 1.6 \times 10^4 \text{k}\Omega$ is the initial value of the state of the memristor, μ_{V} is the average mobility of oxygen deficiencies, $D = 10\text{nm}$ is the thickness of the film of the memristor device, the initial value of the system $(x_0, y_0, z_0) = (0.3, 0.1, 0.2)$ (the same as below), the variable step-size fourth-order Runge-Kutta algorithm is used to simulate the above two systems. Their attractor phase diagram is shown in **Figure 11**.

Both system Eq. (27) and system Eq. (28) present a chaotic attractor with two scrolls in the plane, one scroll trajectory of the Lorenz chaotic system is relatively concentrated, and its surrounding trajectories are attracted in the region and spread around, while the distribution law of the two scroll trajectories of the Lorenz memristor chaotic system is not very much different from each other, and maintains the basic structure of the Lorenz chaotic attractor. This is consistent with the rules in the above-proposed method of constructing memristor chaotic systems, which also shows the correctness of the method of constructing memristor chaotic systems proposed in the text, i.e., without changing the structure of the equation of state of the

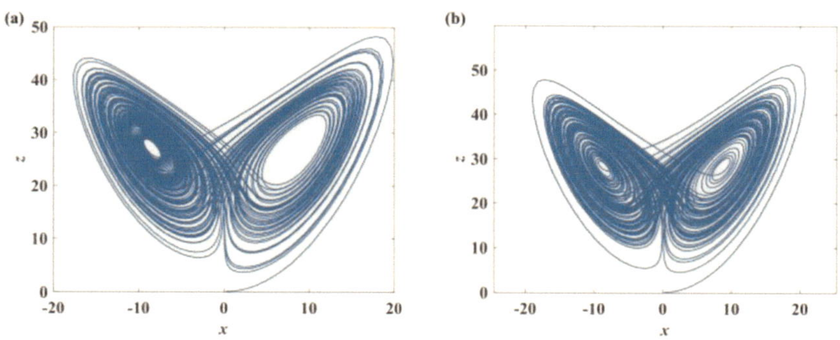

Figure 11.
Chaotic attractors in the $x - y$ plane. (a) Lorenz chaotic system. (b) Lorenz memristor chaotic system.

original system, it can also basically maintain the structure of the attractor phase diagram of the original system, whereas the degree of chaos and the robustness of the system need to be determined from several performance indicators.

To investigate the specific effect of the HP memristor in the original system, the control parameter $\gamma \in [0, 5]$ is set, and the bifurcation diagrams of the Lorenz chaotic system and the Lorenz memristor chaotic system are plotted respectively, as shown in **Figure 12(a)**. The bifurcation diagram of Lorenz chaotic system is divided into three parts: periodic state, chaotic state, and non-chaotic state, when $\gamma = 3.779$, the system is not chaotic, while in the bifurcation diagram of Lorenz memristor chaotic system, although there is no significant change in the region of the periodic state, but the region of the chaotic state is increased in **Figure 12(b)**. This shows that the HP memristor can effectively extend the parameter interval of the original Lorenz chaotic system.

To study and compare the degree of chaos of the two systems, the LLE method and Multi-scale permutation entropy (MPE) [47] are used to examine the degree of chaos and the complexity of the sequences of the two systems, respectively. As shown in **Figure 13**, the LLE of the Lorenz memristor chaotic system is significantly larger compared to the Lorenz chaotic system, when $\gamma = 3.5$, the LLE of the Lorenz chaotic system rapidly decreases to 0, and in the subsequent parameter control, the LLE of the system is constantly less than 0, at which time the system is in a non-chaotic state, while the LLE of the Lorenz memristor chaotic system is continuously maintained at around 1, which remains consistent with the phenomenon of the bifurcation diagram. In the same case, the MPE of the system exhibits the same trend as the LLE, which demonstrates that the generalized HP memristor model proposed in this paper can

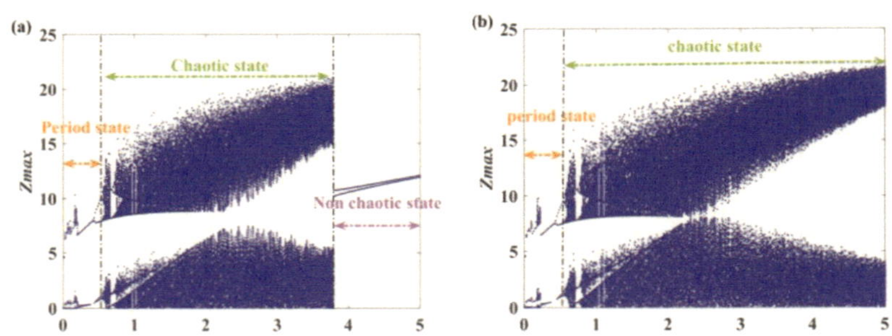

Figure 12.
Bifurcation of the system with $\gamma \in [0, 5]$. (a) Lorenz chaotic system. (b) Lorenz memristor chaotic system.

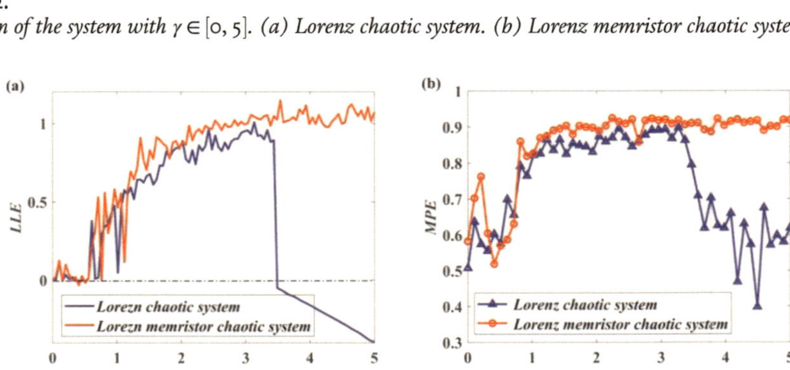

Figure 13.
LLE and MPE of two systems (a) LLE; (b) MPE.

	Chaos range of γ	Larger LLE ratio	Larger MPE ratio
Lorenz	3.249	13%	5%
Lorenz memristor	4.470	87%	95%

Table 4.
Comparison of Lorenz chaotic system and Lorenz memristor chaotic system.

effectively expand the parameter range of the system and strengthen the degree of chaos of the system without changing the structure of the original system equations and the attractor structure.

In **Table 4**, the chaos range, larger LLE ratio and larger MPE ratio under the parameter γ for the Lorenz memristor chaotic system discussed in this section and the classical Lorenz chaotic system are given, demonstrating that the Lorenz memristor chaotic system has a larger chaos range and a stronger degree of chaos compared to the classical system.

3. Conclusion

Memristor is a kind of component with great development possibility, which has the characteristics of small size, low power consumption, and memory function, and has high application value in chaotic circuits. In this paper, the nonlinear characteristics and the mechanism of action inside the memristor are investigated through theoretical analysis. Firstly, two types of generalized memristor models are proposed. On the one hand, based on sampling memristor-capacitor circuits, a generalized voltage-controlled discrete memristor model is derived through the intrinsic relationship between the memristor and the capacitor. On the other hand, under the definition of charge-flux in the HP memristor model, considering the operating initial conditions of the memristor device and the film boundary movement conditions, the interdependence between charge, magnetic flux, and memristor is analyzed in detail, and a generalized magnetically controlled titanium dioxide memristor model is deduced. Secondly, the above two types of memristor models are applied to two classical chaotic systems, and the construction methods and application rules of the new systems are elaborated in detail. Then, the dynamical behaviors of the original system and the memristor system are compared and analyzed by various dynamical analysis methods to verify the correctness and applicability of the proposed generalized memristor model. Finally, the experimental results show that the generalized memristor model proposed in this paper can effectively strengthen the chaotic characteristics and robustness of the system, which provides an effective theoretical basis and experimental foundation for applications in the field of confidential communication.

Acknowledgements

This work was supported in part by the National Natural Science Foundation of China under Grant 61971109, in part by the GF Science and Technology Special Innovation Zone Project, and in part by the Fundamental Research Funds of Central Universities under Grant 2672018ZYGX2018J009.

Conflict of interest

The authors declare that they have no known competing financial interests or personal relationships that could have appeared to influence the work reported in this paper.

Author details

Xiangliang Xu[1*], Tianxian Zhang[1*], Yongqi Zhao[1], Wanying Dai[2], Si Wu[3] and Guodong Li[3]

1 School of Information and Communication Engineering, University of Electronic Science and Technology of China, Chengdu, China

2 School of Cyber Science and Engineering, Sichuan University, Chengdu, P.R. China

3 School of Mathematics and Computational Science, Guilin University of Electronic Technology Guilin, Guangxi, People's Republic of China

*Address all correspondence to: xd160622@126.com and txzhang@uestc.edu.cn

References

[1] Chua L. Memristor-The missing circuit element[J]. IEEE Transactions on Circuit Theory. 1971;**18**(5):507-519. DOI: 10.1109/TCT.1971.1083337

[2] Strukov DB, Snider GS, Stewart DR, et al. The missing memristor found[J]. Nature. 2008;**453**(7191):80-83. DOI: 10.1038/nature06932

[3] Wen SP, Wei HQ, Zheng Y, et al. Memristor-based design of sparse compact convolutional neural network [J]. IEEE Transactions on Network Science and Engineering. 2020;**7**: 1431-1440. DOI: 10.1109/tnse.2019.2934357

[4] Cai F, Kumar S, Van Vaerenbergh T, et al. Power-efficient combinatorial optimization using intrinsic noise in memristor Hopfield neural networks[J]. Nature Electronics. 2020;**3**(7):409-418. DOI: 10.1038/s41928-020-0436-6

[5] Dai WY, Xu XL, Song XM, et al. Audio encryption algorithm based on Chen memristor chaotic system[J]. Symmetry. 2021;**14**:17. DOI: 10.3390/sym14010017

[6] Wang L, Jiang S, Ge MF, et al. Finite-/fixed-time synchronization of memristor chaotic systems and image encryption application[J]. IEEE Transactions on Circuits and Systems I: Regular Papers. 2021;**68**(12):4957-4969. DOI: 10.1109/tcsi.2021.3121555

[7] Akgul A, Cimen ME, Moroz IM, et al. The modeling of memcapacitor oscillator motion with ANN and its nonlinear control application[M]. In: Mem-Elements for Neuromorphic Circuits with Artificial Intelligence Applications. Amsterdam: Academic Press; 2021. pp. 99-123. DOI: 10.1016/b978-0-12-821184-7.00013-x

[8] Biolek Z, Biolek D, Biolkova V. SPICE model of memristor with nonlinear dopant drift[J]. Radioengineering. 2009; **18**(2):210-214

[9] Wu HG, Bao BC, Chen M. Threshold flux-controlled memristor model and its equivalent circuit implementation[J]. Chinese Physics B. 2014;**23**(11):118401. DOI: 10.1088/1674-1056/23/11/118401

[10] Muthuswamy B. Implementing memristor based chaotic circuits[J]. International Journal of Bifurcation and Chaos. 2010;**20**(05):1335-1350. DOI: 10.1142/s0218127410026514

[11] Hu X, Chen G, Duan S, et al. A memristor-based chaotic system with boundary conditions[M]. In: Memristor Networks. Cham: Springer; 2014. pp. 351-364. DOI: 10.1007/978-3-319-02630-5_16

[12] Sun J, Zhao X, Fang J, et al. Autonomous memristor chaotic systems of infinite chaotic attractors and circuitry realization[J]. Nonlinear Dynamics. 2018;**94**:2879-2887. DOI: 10.1007/s11071-018-4531-4

[13] Li H, Yang Y, Li W, et al. Extremely rich dynamics in a memristor-based chaotic system[J]. The European Physical Journal Plus. 2020;**135**:1-18. DOI: 10.1140/epjp/s13360-020-00569-4

[14] Xu B, Wang G, Iu HHC, et al. A memristor–meminductor-based chaotic system with abundant dynamical behaviors[J]. Nonlinear Dynamics. 2019; **96**:765-788. DOI: 10.1007/s11071-019-04820-1

[15] Lai Q, Wan Z, Kengne LK, et al. Two-memristor-based chaotic system with infinite coexisting attractors[J].

IEEE Transactions on Circuits and Systems II: Express Briefs. 2020;**68**(6): 2197-2201. DOI: 10.1109/tcsii.2020.3044096

[16] Li G, Xu X, Zhong H. A image encryption algorithm based on coexisting multi-attractors in a spherical chaotic system[J]. Multimedia Tools and Applications. 2022;**81**(22):32005-32031. DOI: 10.1007/s11042-022-12853-9

[17] Li H, Hua Z, Bao H, et al. Two-dimensional memristive hyperchaotic maps and application in secure communication[J]. IEEE Transactions on Industrial Electronics. 2020;**68**(10): 9931-9940. DOI: 10.1109/tie.2020.3022539

[18] Yang N, Xu C, Wu C, et al. Modeling and analysis of a fractional-order generalized memristor-based chaotic system and circuit implementation[J]. International Journal of Bifurcation and Chaos. 2017;**27**(13):1750199. DOI: 10.1142/s0218127417501991

[19] Bo-Cheng B, Jian-Ping X, Zhong L. Initial state dependent dynamical behaviors in a memristor based chaotic circuit[J]. Chinese Physics Letters. 2010; **27**(7):070504. DOI: 10.1088/0256-307x/27/7/070504

[20] Akgül A, Rajagopal K, Durdu A, et al. A simple fractional-order chaotic system based on memristor and memcapacitor and its synchronization application[J]. Chaos, Solitons & Fractals. 2021;**152**:111306. DOI: 10.1016/j.chaos.2021.111306

[21] Xie W, Wang C, Lin H. A fractional-order multi-stable locally active memristor and its chaotic system with transient transition, state jump[J]. Nonlinear Dynamics. 2021;**104**(4): 4523-4541. DOI: 10.1007/s11071-021-06476-2

[22] Deng Y, Li Y. A memristive conservative chaotic circuit consisting of a memristor and a capacitor[J]. Chaos: An Interdisciplinary Journal of Nonlinear Science. 2020;**30**(1):013120. DOI: 10.1063/1.5128384

[23] Yuan F, Jin Y, Li Y. Self-reproducing chaos and bursting oscillation analysis in a meminductor-based conservative system[J]. Chaos: An Interdisciplinary Journal of Nonlinear Science. 2020;**30**(5):053127. DOI: 10.1063/5.0008313

[24] Yuan F, Wang G, Wang X. Extreme multistability in a memristor-based multi-scroll hyper-chaotic system[J]. Chaos: An Interdisciplinary Journal of Nonlinear Science. 2016;**26**(7):073107. DOI: 10.1063/1.4958296

[25] Alombah NH, Fotsin H, Ngouonkadi EBM, et al. Dynamics, analysis and implementation of a multi-scroll memristor-based chaotic circuit[J]. International Journal of Bifurcation and Chaos. 2016;**26**(08):1650128. DOI: 10.1142/S0218127416501285

[26] Li H, Wang L, Duan S. A memristor-based scroll chaotic system—Design, analysis and circuit implementation[J]. International Journal of Bifurcation and Chaos. 2014;**24**(07):1450099. DOI: 10.1142/s0218127414500990

[27] Yan D, Wang L, Duan S, et al. Chaotic attractors generated by a memristor-based chaotic system and Julia fractal[J]. Chaos, Solitons & Fractals. 2021;**146**:110773. DOI: 10.1016/j.chaos.2021.110773

[28] Xu XL, Li GD, Dai WY, et al. Multi-direction chain and grid chaotic system based on Julia fractal[J]. Fractals. 2021; **29**(08):2150245. DOI: 10.1142/s0218348x21502455

[29] Wei C, Li G, Xu X. Design of a new dimension-changeable hyperchaotic model based on discrete memristor[J]. Symmetry. 2022;**14**(5):1019. DOI: 10.3390/sym14051019

[30] Bao BC, Li H, Wu H, et al. Hyperchaos in a second-order discrete memristor-based map model[J]. Electronics Letters. 2020;**56**(15):769-770. DOI: 10.1049/el.2020.1172

[31] Peng Y, Sun K, He S. A discrete memristor model and its application in Hénon map[J]. Chaos, Solitons & Fractals. 2020;**137**:109873. DOI: 10.1016/j.chaos.2020.109873

[32] Li G, Zhong H, Xu W, et al. Two modified chaotic maps based on discrete memristor model[J]. Symmetry. 2022;**14**(4):800. DOI: 10.3390/sym14040800

[33] Peng Y, He S, Sun K. Chaos in the discrete memristor-based system with fractional-order difference[J]. Results in Physics. 2021;**24**:104106. DOI: 10.1016/j.rinp.2021.104106

[34] Bao H, Hua Z, Li H, et al. Discrete memristor hyperchaotic maps[J]. IEEE Transactions on Circuits and Systems I: Regular Papers. 2021;**68**(11):4534-4544. DOI: 10.1109/tcsi.2021.3082895

[35] Zhong H, Li G, Xu X. A generic voltage-controlled discrete memristor model and its application in chaotic map [J]. Chaos, Solitons & Fractals. 2022;**161**:112389. DOI: 10.1016/j.chaos.2022.112389

[36] Yao P, Wu H, Gao B, et al. Fully hardware-implemented memristor convolutional neural network[J]. Nature. 2020;**577**(7792):641-646. DOI: 10.1038/s41586-020-1942-4

[37] Li C, Belkin D, Li Y, et al. Efficient and self-adaptive in-situ learning in multilayer memristor neural networks[J]. Nature Communications. 2018;**9**(1):2385. DOI: 10.1038/s41467-018-04484-2

[38] Lin H, Wang C, Hong Q, et al. A multi-stable memristor and its application in a neural network[J]. IEEE Transactions on Circuits and Systems II: Express Briefs. 2020;**67**(12):3472-3476. DOI: 10.1109/tcsii.2020.3000492

[39] Zhang Y, Wang X, Friedman EG. Memristor-based circuit design for multilayer neural networks[J]. IEEE Transactions on Circuits and Systems I: Regular Papers. 2017;**65**(2):677-686. DOI: 10.1109/tcsi.2017.2729787

[40] Wang L, Drakakis E, Duan S, et al. Memristor model and its application for chaos generation[J]. International Journal of Bifurcation and Chaos. 2012;**22**(08):1250205. DOI: 10.1142/S0218127412502057

[41] Batas D, Fiedler H. A memristor SPICE implementation and a new approach for magnetic flux-controlled memristor modeling[J]. IEEE Transactions on Nanotechnology. 2010;**10**(2):250-255. DOI: 10.1109/tnano.2009.2038051

[42] Chua L. If it's pinched it'sa memristor[J]. Semiconductor Science and Technology. 2014;**29**(10):104001. DOI: 10.1088/0268-1242/29/10/104001

[43] He S, Sun K, Peng Y, et al. Modeling of discrete fracmemristor and its application[J]. AIP Advances. 2020;**10**(1):015332. DOI: 10.1063/1.5134981

[44] Long-Xiang F, Shao-Bo H, Hui-Hai W, et al. Simulink modeling and dynamic characteristics of discrete memristor chaotic system[J]. Acta Physica Sinica. 2022;**71**(3):030501. DOI: 10.7498/aps.71.20211549

[45] Adhikari SP, Sah MP, Kim H, et al. Three fingerprints of memristor[J]. IEEE Transactions on Circuits and Systems I: Regular Papers. 2013;**60**(11):3008-3021. DOI: 10.1109/tcsi.2013.2256171

[46] Hénon M. A two-dimensional mapping with a strange attractor[J]. In: The Theory of Chaotic Attractors. New York: Springer; 2004. pp. 94-102. DOI: 10.1007/978-0-387-21830-4_8

[47] Morabito FC, Labate D, Foresta FL, et al. Multivariate multi-scale permutation entropy for complexity analysis of Alzheimer's disease EEG[J]. Entropy. 2012;**14**(7):1186-1202. DOI: 10.3390/e14071186

Chapter 7

Bipolar Resistive Memory with Functional Duality-Non Volatile Emerging Memory and Nano Biosensors

Sourav Roy

Abstract

Resistive Memory in recent year has emerged as a potential candidate in the field of Non Volatile Memory to solve the existing problems with FLASH. The operation novelty of ReRAM helps to evolve it from storage device to an effective ultra sensitive biomarker with a very simple structure and fabrication process steps. Basically as ReRAM is MIM capacitor like structure so for store data in terms of charge like DRAM is feasible effectively and at the same time capacitor as we know can be excellent for bioanalyte detection. So with same structure two purpose can be solved. Also we can see in this chapter that the biosensors with ReRAM will detect on Current -Voltage sampling method which is more efficient to detect with low sample volume. This chapter will give the readers a brief idea about the work done and ongoing research on Resistive memory as Non Volatile Memory as well as its potentiality as Biosensor.

Keywords: ReRAM, MIM, NVM, biosensor, bioanalytes, IV, CV, HRS, LRS, TMO

1. Introduction

Modern era is the era of electronic gadgets and semiconductor memory is not only the integrated part of the modern electronics devices rather it acts as a heart of the device. From a longer period of time different research groups across the world in both industries and academics are working on various semiconductor memories and try to find the way out for better performances. The demand of development of the semiconductor memories are boosted due to enhancement usability of personal electronics appliances as well as amusement electronic gadgets such as laptops, palmtops, digital camera, tablet PCs, various kinds of smartphones, etc. [1]. This memory based electronic gadgets also play an important role in medical market. As the days passes the dramatic development in all these electronic gadgets put a demand of more powerful memory devices in front of researchers. On the basis of operation principle the semiconductor memory is classified into two major groups: volatile and non-volatile memory (NVM) [2]. Volatile memories are such kinds of memories which require an external power source to retain data. Static random access memory (SRAM) and dynamic random access memory (DRAM) are the two existing volatile memory in arket. On the other hand NVM are those kind of semiconductor memories which do not need any external power source for retaining their stored data. Some popular NVM leading in semiconductor memory market are phase change memory (PCM), FLASH memory, ferroelectric random access memory (FeRAM) and magnetic random access memory (MRAM) etc. Now all these memory devices have several advantages as well as disadvantages. For example DRAM is very high capacitive and higher density memory structure but it needs a refreshment operation in regular intervals of few milliseconds. On the other hand SRAMs have the ability of faster operation and it possesses a large numbers of memory cells but it has lower capacity than DRAMs. Among them the most leading and popular NVM that commercially rule the current semiconductor memory market is FLASH memory. The FLASH memory has two basic advantages for its rapid growth and applications [3]; one is its miniaturized structure and reliability property. Another one is its non volatility characteristics. But still the scaling issue and some operational shortcomings hinder FLASH memories to lead the market for longer periods of time. Additionally the demand of new NVM that can overcome the drawbacks of FLASH increases day by day. Some important key pitfalls of FLASH are limited endurance ($\sim 10^5$ to 10^6 cycles), small write-erase (W/E) time (1 ms to 1 μs) and higher operation voltage [4, 5]. But above all the serious bottleneck issue for FLASH is its scaling limit (cannot be scaled below 20 nm). So the requirement of emerging NVM which can capable to replace the FLASH is necessary. The market expectation is that this new kind of NVM can able to show longer endurance ($>10^6$ cycles) with faster switching speed (few ns) and stable data retain ability property ($\sim$10 years @ 85°C). Additionally this new kind of memory devices should have smaller feature size (<20 nm) and show lesser power consumption property and must be CMOS compatible. In this respect International Road Map for Semiconductors (ITRS) had proposed a guideline in 2015 where a brief comparison with emerging NVM technologies had a very good comparison with FLASH and showed which are the emerging NVM that have the potentiality to take over the lead in future NVM market from FLASH [6]. The comparisons are shown in **Table 1**.

Basically Resistive Memory devices are structure wise a simple metal-insulator-metal capacitor based devices, where a high-k dielectric insulator layer mostly made of metal oxide (MOs) or transitional metal oxides (TMOs) is sandwiched between two terminal metal electrodes. Depending on the types of material of the top metal

		Base line technologies				Prototypical technologies			Emerging
		DRAM	SRAM	Flash		FeRAM	STT-MRAM	PCM	Redox FeRAM
				NOR	NAND				
Cell elements		1 T1C	6 T	1 T		1 T1C	1(2)T1R	1 T(D)1R	1 T(D)1R
Feature size, nm	2013	36	45	90	22	180	65	45	D:9 B: <5
	2026	9	10	25	8	65	16	8	
Cell area	2013	6 F^2	140 F^2	10 F^2	4 F^2	22 F^2	20 F^2	4F^2	4F^2
	2026	4 F^2	140 F^2	10 F2	4 F^2	12 F^2	8F^2	4F^2	
Read time	2013	<10 ns	0.2 ns	15 ns	0.1 ms	40 ns	35 ns	12 ns	D: <50 ns
	2026	< 10 ns	70 ps	8 ns	0.1 ms	<20 ns	<10 ns	<10 ns	B: <10 ns

		Base line technologies				Prototypical technologies			Emerging
		DRAM	SRAM	Flash		FeRAM	STT-MRAM	PCM	Redox FeRAM
				NOR	NAND				
W/E time	2013	<10 ns	0.2 ns	1 μs/1 ms	1/0.1 ms	65 ns	35 ns	100 ns	0.3 ns
	2026	<10 ns	70 ps	1 μs/1 ms	1/0.1 m s	<10 ns	<1 ns	<50 ns	
Retention time	2013	64 ms	—	10 y	10 y	10y	>10 y	>10y	10 y
	2026	64 ms	—	10 y	10 y	10y	>10 y	>10y	
Write cycles	2013	>1E16	>1E16	1E5	1E14	1E14	>1E12	1E9	D: 1E12 B: 1E16
	2026	>1E16	>1E16	1E5	5E3	>1E15	>1E15	1E9	
Write voltage (V)	2013	2.5	1	10	15	1.3–3.3	1.8	3	< 0.5
	2026	1.5	0.7	9	15	0.7–1.5	<1	<3	
Read voltage (V)	2013	1.8	1	1.8	1.8	1.3–3.3	1.8	1.2	0.15
	2026	1.5	0.7	1	1	0.7–1.5	<1	<1	
Write energy (J/bit)	2013	4E-15	5E-16	1E-10	>2E-16	3E-14	2.5E-12	6E-12	D: 1E-13, B: 1E-17
	2023	2E-15	3E-17	IE-11	>2E-17	7E-15	1.5E-13	1E-15	

Table 1.
Comparison of characteristics of DRAM,SRAM,FLASH with other emerging NVM published in international technology roadmap for semiconductor (ITRS), emerging research devices (ERD) chapter, 2015 [6].

electrodes the Resistive Memory are categorized into 2 types- (a) **Conductive Bridging Random Access Memory (CBRAM)** and (b) **Resistive Random Access Memory (ReRAM)**.

On the other hand, resistive switching mechanism in MO/TMOs-based structure have been reported in various articles already [7–11]. But exploring the features of Resistive memory not as a storage device but as biosensor is very rare [12–17]. Now along with memory characterization some bioanalyte detection tests also are carried out because these biological compounds are most important for maintaining a healthy body. As resistive switching is based on reduction-oxidation (redox) process, also many of the bioanalytes detection are based on redox process where the one of the common by product is H_2O_2, so the switching mechanism also can be justified through H2O2 sensing. Hence the interest on more work on RRAM as biosensor is hoped to come in upcoming time.

Hence, different bio-analyte, namely Creatinine, Urea, Sarcosine, LoXL2, Tributyrin, etc. detected by 2D Through Via Hole (TVH) and 3D cross-point (CP) structure is a completely new concept demonstrated in this chapter.

In this chapter we will do a basic review on ReRAM, CBRAM, 3D ReRAM or Cross-Point Memory and ReRAM as Biosensor in the following sections.

2. Resistive memory as a promising alternative of FLASH

Conventional semiconductor memories such as DRAM, FLASH have already reached to some limitations from technological point of view and so a replacement alternative is emerged as market demand. Now some emerging candidates in the field of NVM which are going to be considered as alternative of FLASH and considered as

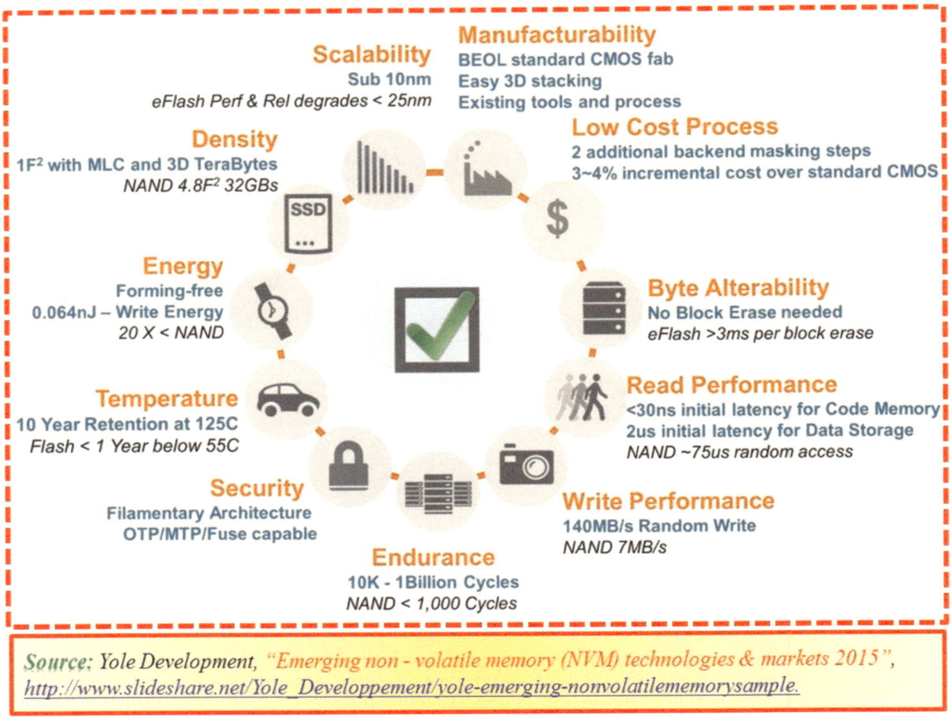

Figure 1.
Advantage of resistive memory over FLASH memory [22].

next generation NVM lead the semiconductor market are PCM, FeRAM, MRAM, and resistive random-access memory (RRAM). Now among all these memories RRAM has emerged as one best option due to several advantageous potentials such as it has a simple MIM structure, CMOS compatibility, excellent scalability potential, and long program/erase (P/E) endurance [18–21]. From technological point of view RRAM is a promising candidate to replace FLASH according to ITRS report published in 2015 [6] as described in **Table 1**. According to this report, where FLASH cannot be scaled down below 20 nm there RRAM has demonstrated to be scaled down below 10 nm and are expected to be scaled down to below 5 nm by 2026. The processing cost is also very less in case of RRAM owing to its simple structure. There is no need of block erase like FLASH (block erase needed for every 3 ms interval). RRAM has excellent latency time (<30 ns) over NAND FLASH (~75 μs) and superior write performance (140 MBPS for RRAM compare to 7 MBPS for NAND FLASH). Endurance for RRAM also very high (>10^{12} [22] and > 10^{16} is expected to achieve). RRAM has ability of longer data retention in robust condition (~10 years at 125°C temperature). Also from other properties like high memory density, low power consumption, low energy, better security are some key advantages of RRAM that make it's a promising candidate for future leader in semiconductor memory field. **Figure 1** shows the reported data that prove RRAMs' superiority over conventional FLASH.

2.1 Resistive RAM basics

Normally the metal oxides (MOs) or transitional metal oxides (TMOs) are acting as insulator at normal state. But if an electric field is applied across them then at certain field the oxide breaks down and a conducting nano-filament is formed which decreases the resistance of the device. Almost more than 40 years ago such kind of fact was come to the picture when in 1962 Hickmott observed a resistive switching phenomena in a MIM structure [23]. Since that the research groups across the world became interested to evaluate the operation principle of RRAM and tried to come out with promising features of resistive switching that make it the suitable alternative of FLASH. The insulating material includes binary or multinary oxides, chalcogenides materials and also organic compounds. Now this insulator is sandwiched between two metal electrodes such as W, Pt, Cu, TiN, etc.

As per the different sources and research groups working globally Resistive random-access memory (ReRAM) is based on a simple three-layer structure of a top electrode, switching medium and bottom electrode as depicted in **Figure 2**. The resistance switching mechanism is based on the formation of a filament in the switching material when a voltage is applied between the two electrodes. There are different approaches to implementing ReRAM, based on different switching materials and memory cell organization. Those variables drive significant performance differences depending upon the switching materials being used. In addition to use as multiple-time programmable (MTP), few-time programmable (FTP) and on-time programmable (OTP) non-volatile memory applications.

ReRAM technology also can be used for security applications, where ReRAM is used for secure physical unclonable function (PUF) keys embedded in semiconductors.

Normally when a positive or negative bias is applied to the top electrode (TE) to create the filament and the bottom electrode (BE) is kept as grounded. The insulator or dielectric material between them is termed as switching material (SM).

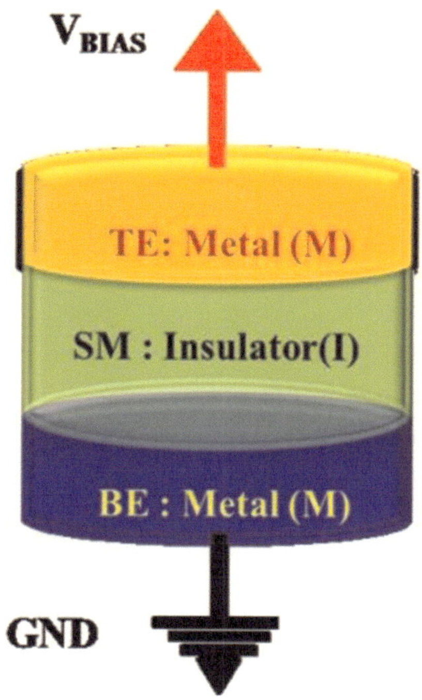

Figure 2.
Simple 2D schematic of resistive memory structure.

Because the resistance switching mechanism is based on an electric field, the ReRAM cell is very stable, capable of withstanding temperature swings from −40°C to 125°C, 1 M+ write cycles and a retention of 10 years at 85°C.

2.2 Resistive switching mechanism

2.2.1 Modes of resistive switching: unipolar and bipolar

From mode of operation point of view resistive switching is classified into two sub category. One is called bipolar switching and another is called Unipolar switching. In a Unipolar switching the direction of current voltage (I-V) hysteresis curve do not depends on the polarity of the applied bias where as in case of bipolar resistive switching mode this phenomenon is just opposite. In bipolar switching the switching direction is highly depends on bias polarity. The clear view of both kinds of resistive switching is described in **Figure 3.**

In most of the publications based on ReRAM or CBRAM are describing bi-polar mechanism and in this article I am also describe about the bi-polar switching only.

2.2.2 Bipolar resistive switching mechanism

Usually the pristine device has a very high insulation property. An initial forming voltage [25] is required to change the initial resistance state (IRS) of the device. Some devices do not require any forming voltage. These are called forming-free devices [26, 27] and these are effective for low power low cost RRAM applications. Now different research groups have developed many physical models to describe the resistive switching phenomena but the exact mechanism occurs inside the device is yet to have a clear conception. In overall point of view the mechanism of resistive switching is described by two established models.

a. **Filamentary type resistive switching** [8] and

b. **Interface mediated resistive switching** [28]

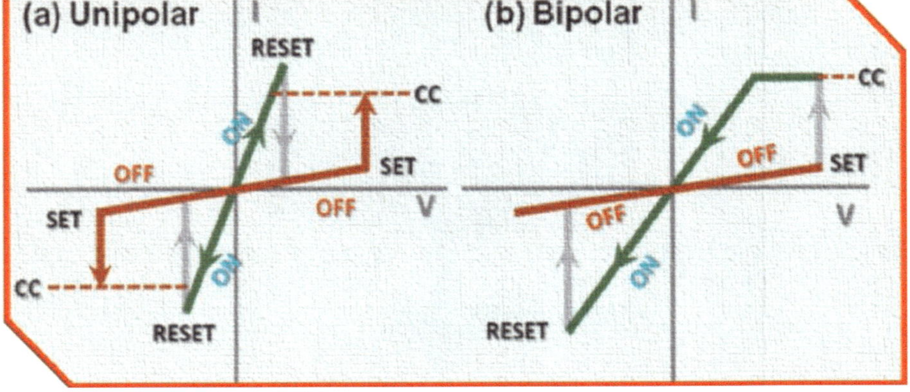

Figure 3.
Typical unipolar and bi-polar resistive switching characteristics of RRAM [24].

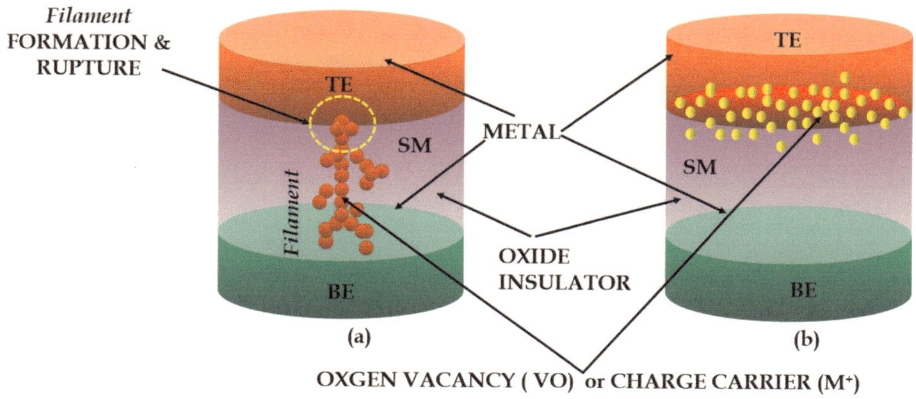

Figure 4.
Typical switching mechanism models of RRAM (a) filamentary type resistive switching model, and (b) Interface mediated resistive switching model.

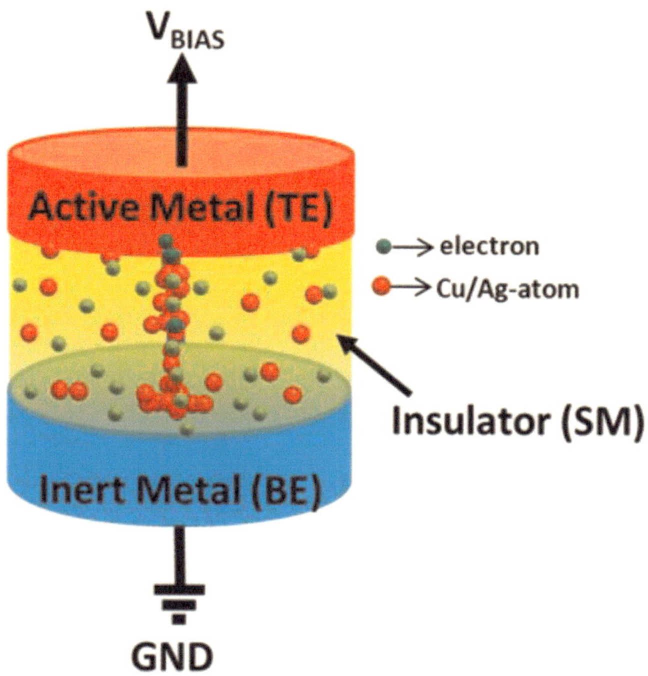

Figure 5.
Typical switching mechanism of cation based resistive switching memory or CBRAM.

These two models shown in **Figure 4** are jointly used to explain the most of the insulating oxides based bipolar resistive switching phenomena [29–31].

In CBRAM structure a thin layer of solid electrolyte or high-κ dielectric material is sandwiched between oxidized anode electrode made of active metals such as Cu, Ag and an inert cathode electrode made of inert metals such as W, Pt, and TiN etc. **Figure 5** describes the CBRAM mechanism using a schematic diagram.

When a positive bias is applied at oxidized electrode the metal ions are oxidized. Now the mobile cations then drifted towards cathode terminal through the dielectric material and form metallic path by connecting two terminal electrodes by a bridge like

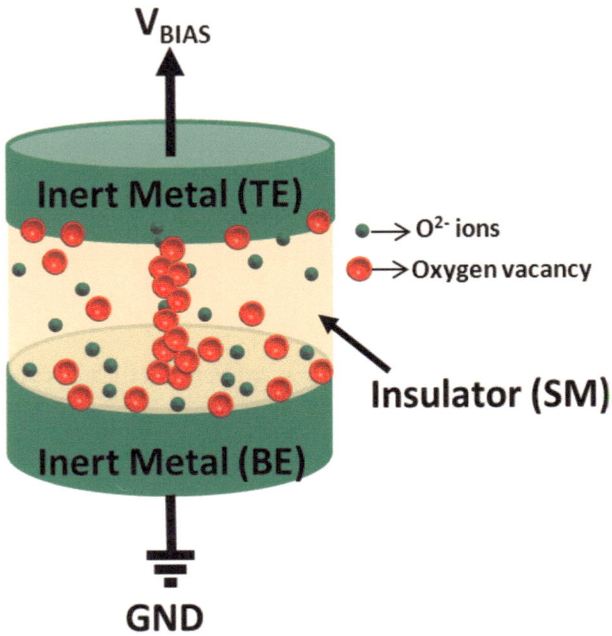

Figure 6.
Typical switching mechanism of anion based resistive switching memory or RRAM.

filamentary path. When the filament is formed the device start to conduct and device resistivity goes to LRS. When the reverse bias is applied again the metal ions drifted towards the active electrode under the influence of applied field and thus a filament dissolution or rupture occurs and the path is broken causing device to switch to HRS.

On the other hand in anion based switching under the influence of applied bias the insulator oxide break down occurs and the O_2-ions are drifted towards the electrode where positive bias is applied.

These results show the generation of oxygen vacancies (VO) inside the insulator material. When there are enough VO causing to form a filamentary path between two terminal electrodes the device starts to switch. Again when the bias with opposite polarity is applied the O2- ions moved in opposite side than the previous case causing the filling up of VO and by this process filament ruptures. This mechanism is described in **Figure 6**.

Although this mechanism is widely proposed by most of the research groups but still the exact mechanism is unfolded. The bias polarity is the main factor that controls the process of oxidation and reduction in this kind of devices [31, 32].

The most accepted CBRAM mechanism was first ever proposed by Prof. M.N. Kozicki [7] in solid electrolyte in 2004 and the same for RRAM was proposed by Prof. R. Waser and M. Aono in 2007 [24].

2.3 Resistive memory materials

To explore the switching mechanism in RRAM devices more elaborately different research groups have studied different switching characteristics by using different switching materials and changing the top and bottom electrodes. Although many materials have been studied but here we are citing few of them showing very

Material	Band gap (Eg) in eV	Dielectric constant (k)	Gibbs free energy (kJ/mole)
Al_2O_3	8.8	9	−1580
$BaTiO_3$	3.42	100–600	−157.4
Gd_2O_3	5.3	17	−1760
a-C	2–5	2.5–6	
GeO_2	0.66	16.2	−518.845
HfO_2	5.8	25	−1010.75
SiO_2	9	3.9	−853.5
$SrTiO_3$	3.2	2000	−1567.77
Ta_2O_5	4.4	22	−760.75
TiO_2	3.2	SO	−923.375
WO_3	3	4.S	−506.875
ZrO_2	5.8	25	−1037.25

Table 2.
Band gap (Eg), dielectric constant (κ) and Gibbs free energy (ΔG) for different metal oxides [34, 35].

promising characteristics and better performances. **Table 2** shows some popular oxides [33–36] used as RRAM dielectrics with their band gap (Eg), dielectric constant (κ) and Gibb's free energy (ΔG).

Among the oxides materials some of the promising candidates are TaOx [10, 18, 22, 37, 38], HfOx [11, 39, 40], and TiOx [21, 41].

Besides that some new oxides are also bring attention to the research groups such as GdOx [12, 42], WOx [43], $BaTiO_3$ [44]. SiOx also have a recent growing interest to the researchers [15, 45–48] because of it has low Gibbs free energy (−853.5 kJ/mole) and high band gap (Eg ~ 9) [34, 35]. Also SiOx is most CMOS compatible material with low cost.

On the other hand, after the proposed mechanism of CBRAM by Kozicki et al. [7], many groups as per reported by Jana et al. [49] giving interest in CBRAM structures. Among the popular materials with which the CBRAM works are carried out mostly are based on solid-electrolyte materials including GeSe [20, 50], GeS2 [51, 52], although there are few reports on high-κ oxides such as Ta_2O_5 [53–57], Al_2O_3 [58–60], TiO_2 [61], and SiOx [47, 62–64]. Some recent reports are based on 2D materials like MoS_2 based CBRAM structures [65, 66].

Among all the switching materials mentioned above Al_2O_3, HfO_2, SiO_2, Ta_2O_5,and TiO_2 are some of the most popular and widely used switching materials owing to their potentiality to generate higher ON/OFF ratio, Longer Program/Erase Endurance cycles, Faster switching speed, and Lower Energy Consumption.

Xiao Liang Hong et al. [67] from NTU Singapore had published a review article in Journal of Material Science in 2018 where they had shortlisted some basic criterion for choosing RRAM materials. Following 3 tables **Tables 3–5** classified RRAM mechanism based on RRAM structure, RRAM top electrode and RRAM switching materials.

It is also observed that combination of Active & Inert electrode results higher endurance and longer data retention where as Graphene & CNT based electrode results ultra dense memory formation.

Type of RRAM Stack Structure	Switching Mechanism Type
a) Symmetric: *Both Terminal electrodes are made with same metal. Like: Pt/HfO_x/Pt; W/ SiO_x/W; TiN/TaO_x/TiN etc.*	Mainly Unipolar Switching are observed.
b) Asymmetric: *Both Terminal electrodes are made with different metal. Like: Pt/HfO_x/TiN; W/SiO_x/TiN; Ir/TaO_x/W etc.*	Mainly Unipolar Switching are observed.

Table 3.
RRAM switching mechanism based on types of RRAM stack structure [67].

Type of RRAM Electrodes	Mechanism
a) Active Metal Electrodes *Metals which can be get oxidized like Ti, Al, Ag, W, Cu, Ni, TiN, TaN, ITO.*	Mainly Bipolar Switching. Switching Types: ECM & VCM. ECM: Electro Chemical Metallization effect. In this case the filament is a metallic filament made of metal cations like Cu, Ag, Ni etc. (CBRAM). VCM: Valance Change Memory effect. Here the filament is fonned by migration of anions or oxygen vacancies. (RRAM).
b) Inert Metal Electrodes *Metals which ne\'er reacts with oxygen such as Pt, Ir, Ru, Au, Pd.*	Mainly Unipolar Switching. Switching Types: VCM.
c) Novel Metal Electrodes *Graphene or CNT.*	Mainly Bipolar Switching.

Table 4.
RRAM switching mechanism based on types of RRAM electrodes [67].

Type of Switching material stack	Mechanism	Advantages
a) Doped Switching layer *If metal oxides doped by me active metals like Ti, Hf Ta, Al or semiconductor materials like Si, Ge*	Dopants diffuse inside the switching materials and create more oxygen vacancies to improve switching and cause fanning free switching	Good data retention, potential of multi bit operation, highly scalability, faster switching speed.
b) Bi-layer / Multilayer Switching material stack *The switching layer will be combination of two metal oxides with one higher and one lower resistivity. Like HfO_x/AlO_x or SiO_x/ TaO_x or GeO_x/TiO_x etc.*	Resistive switching occurs at higher resistivity layer and lower resistivity layer can improve the ON state resistance.	Better uniformity (diffusion of metal ions from the lower resistivity to the higher resistivity layer. This stabilizes the CFs to reduce the randomness of resistive switching (RS), Multi-Level Storage function, Low power operation, lower Reset current.

Table 5.
RRAM switching mechanism based on types of RRAM switching materials [67].

3. 3D resistive memory structure or cross-point memory

To compete the 3D NAND FLASH the RRAM should be upgraded next level for ultra-fast performances. Now for these one of the good options is to go for high density memory applications. For this we need to work on cross-point memory devices. The Cross Bar Architecture (CBA) or Cross-Point (CP) structure is a fundamental structure used in 3D Resistive Switching Memory applications. It consists of a grid-like arrangement of memory cells, with memory cells integrated between word-lines and bit-lines [68] as shown in **Figure** 7.

Seok et al. [68] had reported this cross-point structure allows for a high cell density, as it reduces the cell size to 4F^2, where F represents the feature size. This high-density ReRAM technology can be stacked in 3D, delivering multiple terabytes of storage on a single chip. Its simplicity, stackability below 10 nm and CMOS compatibility enable logic and memory to be integrated onto a single chip at the latest technology node. Designers can put logic, controllers and microprocessors next to memory in the same die, simplifying packaging and increasing performance. ReRAM's simple structure and CMOS compatibility enable any foundry - CMOS or logic - to enter the ReRAM business by licensing CrossBar ReRAM technology for Systems-on-Chip (SoC) or standalone memory devices or embedded cryptographic keys.

However, despite its advantages in terms of cell density, the CBA or CP is not without its challenges. One of the main issues is the occurrence of cross-talk interference between memory cells which called as Sneak Path Current shown by red dotted line in **Figure** 7. This interference is caused by leakage current paths through neighboring cells with low resistances.

Many researchers are hence try to design a high-density memory which can operate at low current operation and deliver a high speed program and erase operation and also reduce the sneak path current. To achieve those requirements, different structures with different materials have been reported by several groups. In CP memory also different SM materials based reports are there among which some of the popular SM are TaOx [69–71], HfOx [72–74] AlOx [13, 75], and SiOx [47, 64].

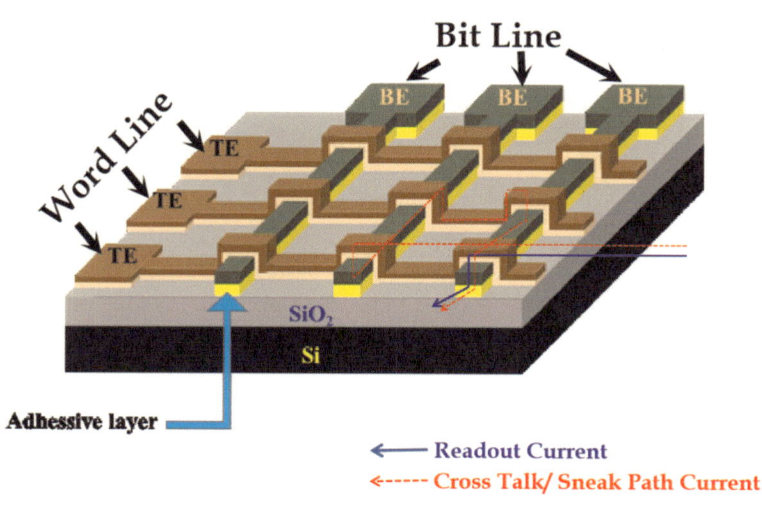

Figure 7.
Typical cross-sectional view of 3 × 3 cross Bar resistive memory device structure.

CP memory can show excellent data storage capability with >3 × 10^6 memory window [74] which make them suitable candidates for high density memory. The cross-point memory show its potentiality for low power VLSI applications with ultra-low operation current of 1 μA [69] and low Set/Reset voltages of ±0.5 V [47]. Also with Cross-bar architecture the longer P/E endurance >10^{11} cycles at low program current ~10 μA [74], excellent data retention >3 hour at robust condition of >125°C temperature [75], high speed operation of 50 ns with low energy per bit of 0.49 pJ [64] can be achievable.

According to the reported data some improvements have been obtained but operation current is still high in most of the cases. To reduce operating current of <1 μA, investigation is needed. In addition, it is also important to detect bioanalyte by using cross-point memory if we want to establish ReRAM has the potentiality to be work as ultra-sensitive biosensor. Although many terminal electrodes are used in CP structure but some recent study has shown if IrO_X will be used as top electrode CP-RRAM can be an excellent biosensor owing to the benefits of Ir for its porous nature shown by the FeSEM image in **Figure 8**, which can help it to act as an excellent sensing membrane for different bioanalytes and help the analytes to reach fast at IrO_X and SM which is the MO interface for redox reaction during switching of the memory device [12, 13].

3.1 Crossbar memory operation

The basic operation of a crossbar-based resistive memory involves applying voltage pulses to the appropriate row and column lines to read or write data. To write data, a voltage pulse is applied across a selected row and column, causing the resistance of the memory cell at the intersection point to switch between high and low states. The resistance state of the memory cell can be sensed by applying a voltage across the row or column and measuring the resulting current.

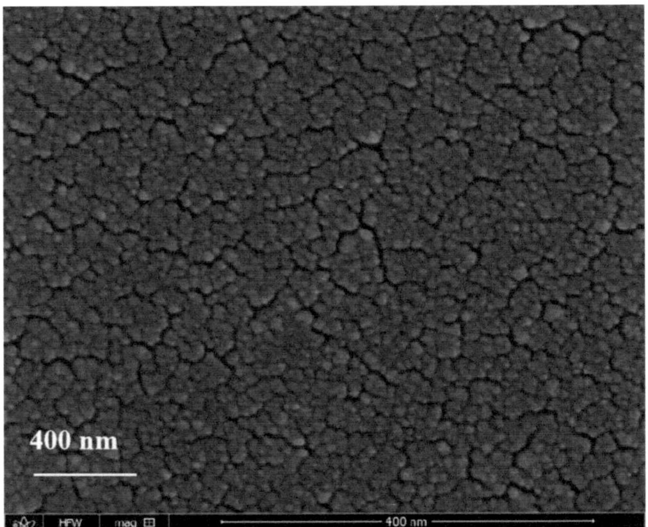

Figure 8.
Field-emission scanning electron microscope (FESEM) image of 20 nm-thick IrO_X layer deposited on Si at 200 k magnification [12].

3.1.1 Read operation

To read data from a memory cell in a crossbar architecture, a voltage is applied across a selected word line (row) and bit line (column). The resulting current through the memory cell is measured to determine its resistance state (HRS or LRS).

3.1.2 Write operation

To write data to a memory cell, a voltage or current pulse is applied across a selected word line and bit line. This pulse induces a physical or chemical change in the insulating layer, causing the resistance of the memory cell to switch between HRS and LRS.

3.2 Crossbar memory advantages and challenges

One of the advantages of the crossbar architecture is its high density. Since each intersection point in the crossbar represents a memory cell, a large number of memory cells can be packed into a small area. This high density makes crossbar-based resistive memory attractive for applications that require large storage capacities.

However, there are several challenges associated with the crossbar architecture. One major challenge is the sneak path problem. When voltage is applied to a selected memory cell, neighboring cells in the same row or column can also be affected due to parasitic currents. This can lead to unintended switching of multiple memory cells, resulting in data corruption. Various techniques such as voltage compliance, current sensing, and advanced algorithms are employed to mitigate the sneak path problem.

Another challenge is the scalability of the crossbar architecture. As the number of memory cells in a crossbar increases, the resistance of the interconnects and the parasitic capacitance also increase, which can degrade the performance and reliability of the memory array. Techniques such as hierarchical and three-dimensional (3D) crossbar architectures are being explored to overcome these scalability limitations.

3.2.1 Sneak path problem

One of the challenges in crossbar-based resistive memory is the sneak path problem. When voltage is applied to a selected memory cell, neighboring cells in the same row or column can be unintentionally affected due to parasitic currents. This can lead to unintended switching and data corruption. Various techniques, such as voltage compliance, current sensing, and advanced algorithms, are employed to mitigate the sneak path problem.

3.2.2 Scalability

The scalability of crossbar-based resistive memory is an important consideration. As the size of the crossbar array increases, there can be challenges in maintaining uniform switching characteristics and managing interconnect resistance and capacitance. Hierarchical and three-dimensional (3D) crossbar architectures are being explored to overcome these scalability limitations.

There are several techniques are employed to mitigate the sneak path problem in crossbar-based resistive memory. Here are some commonly used approaches:

i. **Voltage Compliance:** One approach is to apply voltage compliance limits during write operations. By setting an upper limit on the voltage applied to the selected lines, the parasitic currents in the neighboring cells can be restricted, reducing the likelihood of unintended switching. This technique helps in confining the switching operation to the intended memory cell.

ii. **Current Sensing:** Current sensing techniques can be applied to detect and compensate for the sneak path currents. By measuring the currents flowing through the bit lines during read and write operations, it is possible to identify the contributions from the unintended sneak path currents. By subtracting or compensating for these currents, the desired current corresponding to the targeted memory cell can be accurately determined.

iii. **Advanced Algorithms:** Sophisticated algorithms and error correction codes can be employed to overcome the sneak path problem. These algorithms take into account the behavior of the memory cells and the sneak path currents to reconstruct the intended data accurately. By employing error correction techniques, the impact of sneak path-induced errors can be mitigated.

iv. **Shielding Techniques:** Physical or electrical shielding techniques can be employed to minimize the influence of parasitic currents. For example, placing additional metal layers or insulating materials between the rows and columns can help reduce the coupling between adjacent memory cells and mitigate the sneak path problem.

v. **Selective Access Devices:** Introducing selective access devices, such as transistors or diodes, at each intersection point in the crossbar can help in isolating the memory cell being accessed. These devices act as switches and allow the targeted memory cell to be electrically connected to the selected word and bit lines, while preventing the parasitic currents from affecting neighboring cells.

vi. **Process Optimization:** The fabrication process of the resistive memory can be optimized to minimize the occurrence of sneak path currents. This can involve carefully engineering the materials, interfaces, and dimensions of the memory cells and interconnects to reduce the coupling between adjacent cells.

It's important to note that while these techniques can mitigate the sneak path problem to a certain extent, eliminating it entirely remains a challenge. The choice and effectiveness of the mitigation techniques depend on the specific characteristics and requirements of the resistive memory technology being used. Ongoing research and development efforts are focused on further improving these techniques and developing new approaches to enhance the reliability and performance of crossbar-based resistive memory.

4. Resistive switching memory fabrication process

The fabrication process of resistive random-access memory (RRAM) involves several key steps to create the memory cells and the necessary components. Here are the general steps involved in RRAM fabrication:

a. **Substrate Preparation:** The process begins with the preparation of a suitable substrate, typically a silicon wafer. The substrate is cleaned and prepared to provide a clean and uniform surface for subsequent layers.

b. **Bottom Electrode Deposition:** A thin layer of a conductive material is deposited on the substrate to serve as the bottom electrode of the RRAM cell. Commonly used materials include metals such as titanium (Ti), tungsten (W), or tantalum (Ta).

c. **Insulator Deposition:** A layer of insulating material is deposited on top of the bottom electrode. This insulator layer plays a crucial role in the resistive switching mechanism of RRAM. Common insulator materials used in RRAM include metal oxides such as titanium dioxide (TiO_2), hafnium oxide (HfO_2), or tantalum oxide (TaOx).

d. **Top Electrode Deposition:** A second conductive layer, known as the top electrode, is deposited on top of the insulator layer. The top electrode completes the RRAM cell structure. Similar to the bottom electrode, it is often made of a conductive material like titanium, tungsten, or tantalum for RRAM and Copper or Silver for CBRAM.

e. **Patterning:** Photolithography and etching techniques are used to define the size and shape of the RRAM cells. Photolithography involves the use of photoresist materials and exposure to UV light through a mask to transfer the desired pattern onto the material layers. Subsequent etching processes selectively remove material from specific areas, creating the desired electrode and insulator patterns.

f. **Passivation Layer:** A passivation layer is deposited on top of the RRAM cells to protect them from external contaminants and provide electrical insulation. This layer helps ensure the long-term stability and reliability of the RRAM devices.

g. **Interconnect Formation:** Metal interconnects are formed to connect the RRAM cells to the peripheral circuitry. These interconnects provide the necessary electrical connections for read, write, and control operations. Multiple layers of interconnects may be formed, and dielectric materials are used to isolate and insulate the interconnect layers.

h. **Front-End and Back-End Processes:** Additional steps include the front-end and back-end processes, which involve the integration of transistors, addressing circuits, and other peripheral components. These processes ensure the RRAM devices can be accessed and controlled effectively.

i. **Testing and Packaging:** After fabrication, the RRAM devices undergo rigorous testing to verify their functionality and performance. Once the devices pass the testing phase, they are packaged to protect them from environmental factors and to enable integration into larger electronic systems.

It's important to note that the specific fabrication steps and techniques can vary depending on the RRAM technology, research advancements, and manufacturing processes employed by different manufacturers. The steps mentioned above provide a general overview of the RRAM fabrication process.

A typical fabrication process flow of oxygen vacancy based resistive memory or RRAM having structure of W/SiO_x/TiN reported by Roy et al. [15] is described in **Figure 9** and the same for metallic filament based resistive memory of CBRAM having structure of Cu/Cu-Al alloy/Ta_2O_5/TiN reported by Roy et al. [56] is described in **Figure 10** and the fabrication process of a typical 3 × 3 crossbar RRAM of the author [unpublished] having structure of Ir/SiO_2/W is shown below in **Figure 11** respectively in below.

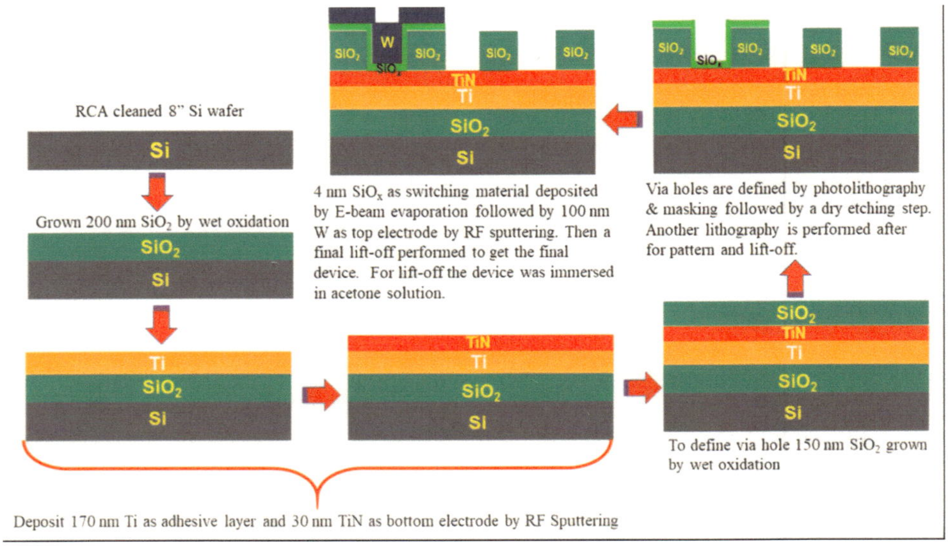

Figure 9.
Typical fabrication process flow step by step of a W/SiO_x/TiN 2D RRAM structure.

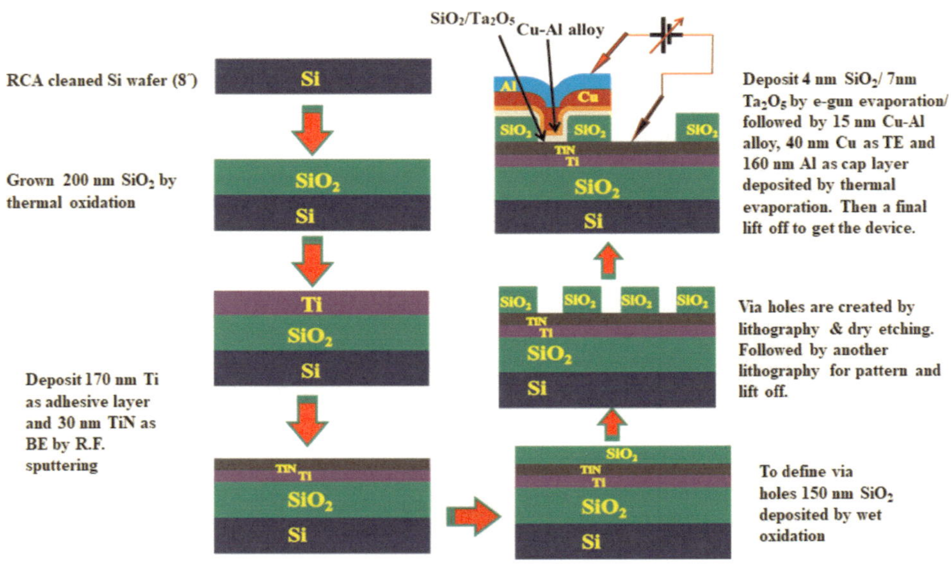

Figure 10.
Typical fabrication process flow step by step of a Cu/Cu-Al alloy/Ta_2O_5/TiN 2D CBRAM structure.

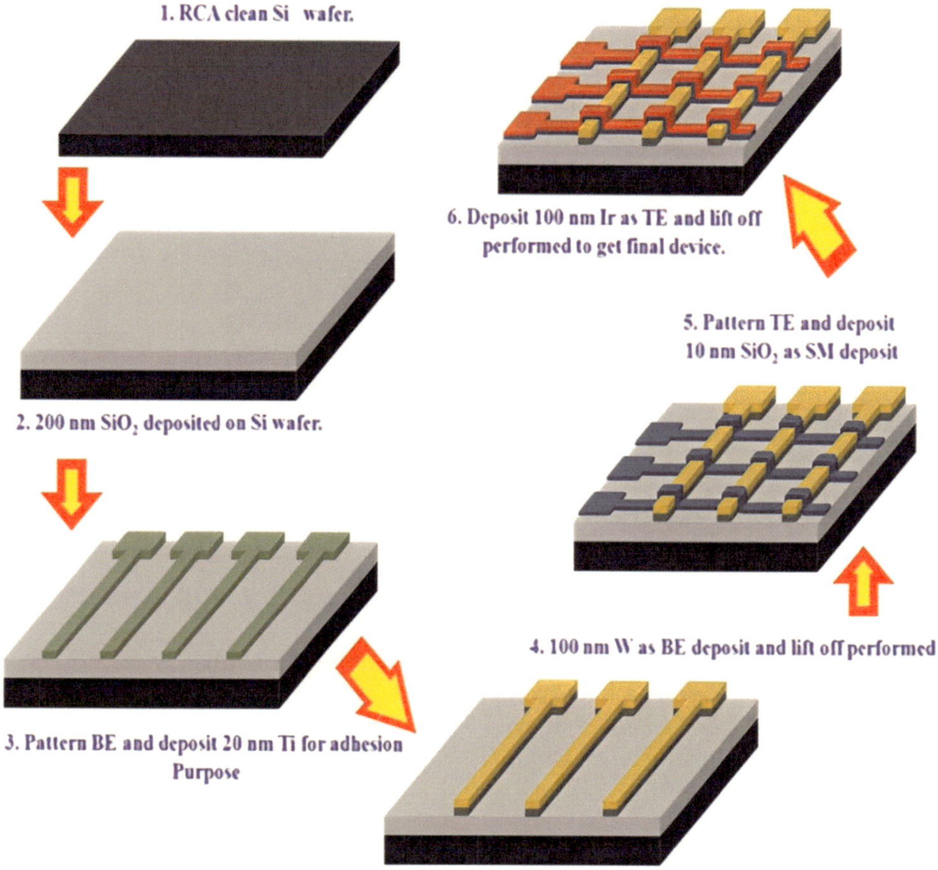

Figure 11.
Typical fabrication process flow step by step of a Ir/SiO_2/W 3D RRAM 3 × 3 Crosspoint structure.

The typical OEM image of a 2D RRAM/CBRAM devices is look like **Figure 12** and a typical SEM image of a single cross point of 3D RRAM is look like **Figure 13** below.

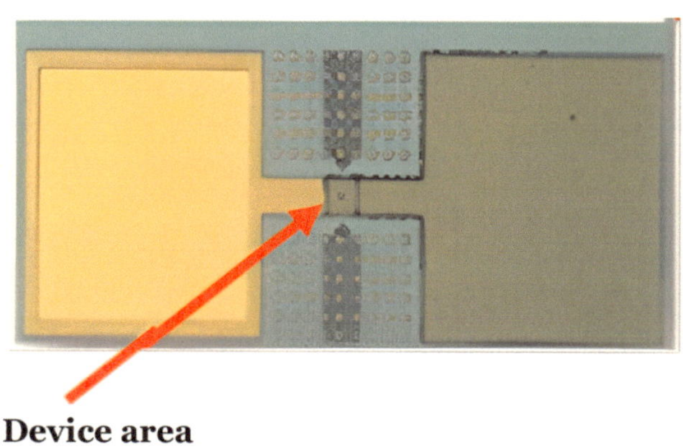

Figure 12.
Typical planer view of OEM image with top (left hand side) and bottom (right hand side) electrode pads of a 2D or through via hole (TVH) resistive memory device.

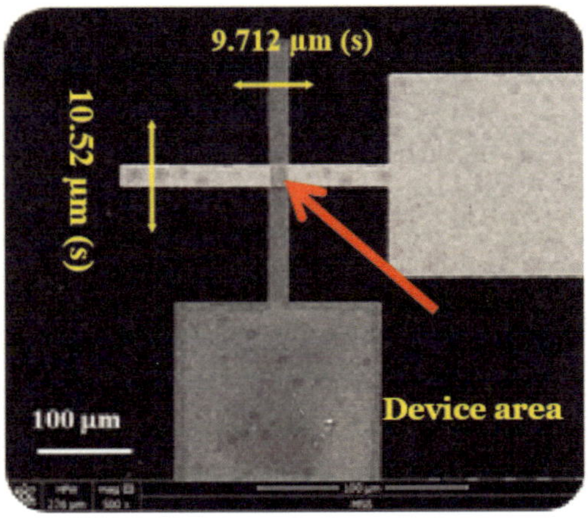

Figure 13.
Typical planer view of SEM image with top (vertical orientation) and bottom (horizontal orientation) electrode pads of a 3D or Crosspoint (single) resistive memory device [13].

5. Typical resistive memory electrical characterization and memory test

As described in Section 2.2 about Resistive Switching mechanism we already know that there are two types of resistive switching mechanism-a) Filamentary type and b) Interface type. **Figures 14** and **15** showed the typical filamentary and interface type switching through pictorial illustration.

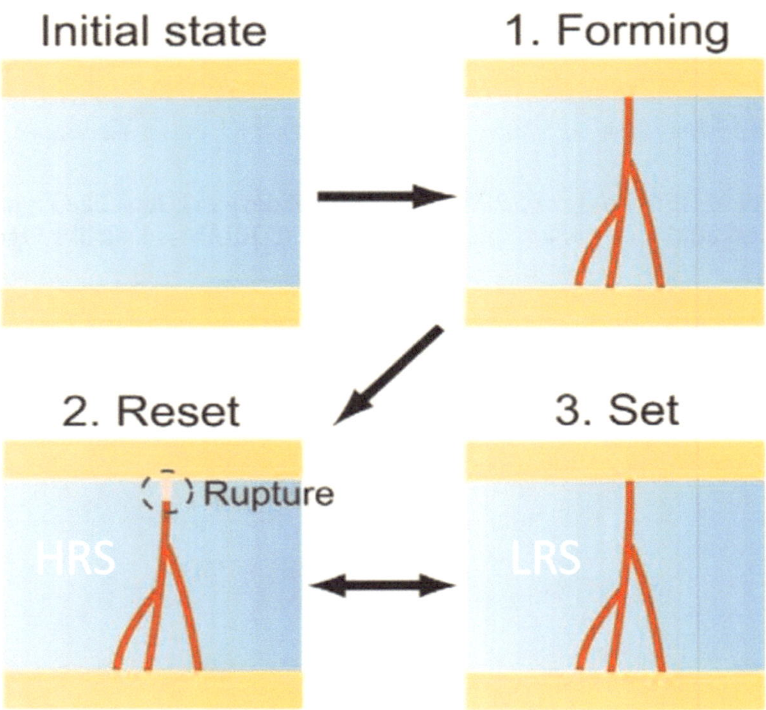

Figure 14.
Typical process diagram of filament formation and rupture of filamentary resistive switching mechanism [8].

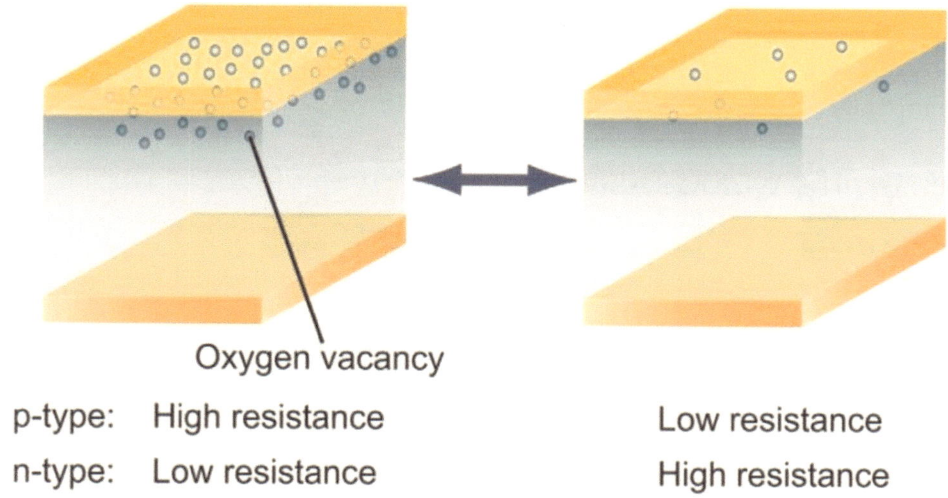

Figure 15.
Typical process diagram of interface type resistive switching mechanism [28].

It is observed from **Figure 14** that in Filamentary type switching when there is no bias is applied the initial state there will be no filamentary path formed inside the insulator layer sandwiched between the terminal electrodes. So the resistance of the insulator will be very high and we called this phase as Initial Resistance State or IRS. Now when first a bias is applied at top electrode a filamentary path will form which may be metallic path for CBRAM devices or may be oxygen vacancy mediated non-metallic filamentary path for RRAM devices. This process is called Formation and the voltage at which the filament forms is called as Formation Voltage. As the filamentary path form so there will be a flow of current through that percolating path inside the insulator switching material and the resistivity of the switching material decreases and that resistance state is known as Low Resistance State or LRS. Now at this point if the Top Electrode bias will be reversed then the filament from the electrode/switching material interfaces start to dissolve or rupture. And immediately the switching material's resistance state again go high and we called that state as High Resistance State or HRS. As this process the complete filament not dissolve and part of it will remain back inside the switching material so the resistance state will not be as high as that during the initial state so HRS is lower than IRS. Now again when the polarity of bias applied at top electrode change then again filament form but this time to form filament lower voltage is required to apply. This voltage is called SET voltage and the voltage required to rupture the filament is called RESET voltage.

The ratio of HRS to LRS is known as Memory Window or Resistance Ratio or ON/OFF Ratio. This ratio is considered to be >10 for an ideal case but as high this ratio will be achieved the data storage capability will be more.

On the other hand for Interface type switching mechanism the transition between HRS to LRS will be mediated by oxygen vacancy concentration at Electrode-Switching material interfaces. If the concentration is high it is denoted as Higher Resistance State or HRS and if the concentration is low then it is called Lower Resistance State or LRS as shown in next page.

It is observed that LRS for filamentary type switching is independent of the device area where as HRS increases as the area decreases. On the other hand for interface type switching both LRS and HRS increases as the device area decreases.

Now in this chapter for memory characterization filamentary effects will be described mainly whereas for biosensor characterization interface type switching is described mainly. For memory characterization usually in general three types of electrical measurements are performed.

5.1 Dual-sweep IV characteristics

In this method a sweep voltage with a particular steps is applied across one of the terminal electrodes where keeping other electrode as grounded, and the current inside the switching material due to formation and rupture of filament are measured. A typical IV characteristics with formation cycle, set operation, reset operation, HRS, LRS, IRS is described in **Figure 16**. The current during formation cycle or the current reach after the device is SET is known as Compliance Current (CC). Whereas current at SET voltage is called Set or Program Current and the same for RESET voltage is called Reset or Erase Current, respectively.

From the figure we can see a current is initially gradually increasing with voltage then at SET voltage it jumps suddenly to a high magnitude indicated by compliance current and then after increasing of voltage current got saturated. Now if voltage is reducing to zero the current initially for a long time remain in saturated level and then suddenly drops to zero. The region in IV curve between 0 V to Set voltage indicated by "1" is called HRS region where as the region between Set voltage to 0 V indicated by "2" is called LRS region.

Now again when the bias polarity is reversed the current will increase from 0 and now the direction of current is reversed. Now at Reset voltage this current suddenly drops to lower value same as HRS state current. This phase the filament got ruptured. As we find both under positive and negative bias polarity there are an exchange of current level with applied voltage between HRS and LRS so this IV sweep also called Bipolar Double Sweep Current Voltage resistive switching characteristics.

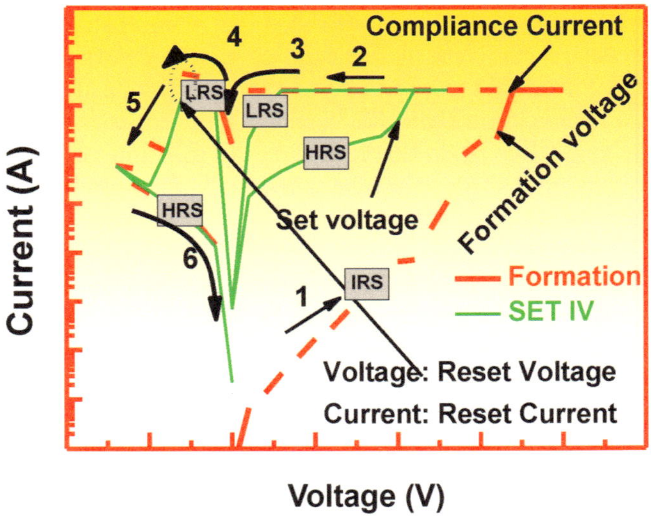

Figure 16.
Typical current–voltage (IV) characteristics of filamentary type of resistive switching mechanism.

5.2 Data retention characteristics

Data retention means under the stress condition how long a memory device can hold the data. To measure this characteristics the device is initially kept in either LRS or HRS region. Then one voltage between 0 V and SET volt has been chosen as Read Voltage and for a longer period of time like 2 or 3 hour or several hours the device resistance state is measured with a time sample having a particular time gap like 2 min or 30 s etc. Then again the state of the device make changes and repeat the same thing for the new state. Thus if it is found that the device can hold the particular value of respective HRS and LRS state at read voltage then it is considered the device has good data retention capability. This measurement can be done at room temperature as well as higher temperature. One the typical data retention author had measured for a Pt/HfO_2/TiN based devices at 100°C is shown in next page (**Figure 17**).

5.3 Endurance characteristics

Endurance characteristics measurement show that how the device can hold the data if a stress is applied continuously to the device inform of AC or DC voltage. Based on the type of stress signal applied across the device the Endurance characteristics is of two types.

5.3.1 DC endurance

In this process a representative DC IV sweep has been carried out and it is checked how long the memory window or ON/OFF ratio can hold the value >10. A typical 3000 repetitive cycles at 300 μA of Cu/Ir/TiNxOy/TiN CBRAM device reported by Dutta et al. [17] has been shown below (**Figure 18**).

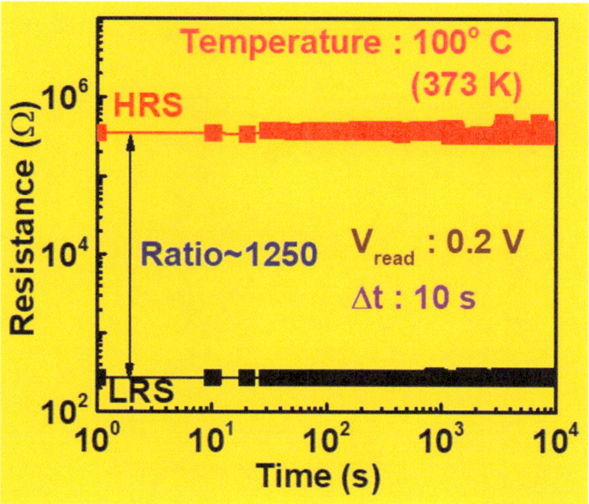

Figure 17.
Typical data retention characteristics of filamentary type of resistive switching mechanism at 100°C temperature [11].

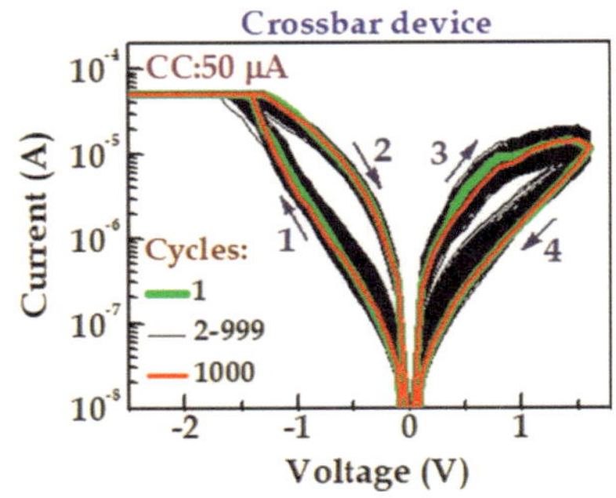

Figure 18.
Typical 1000 representative DC cycles of a Cross-Point array at 50 mA compliance current [17].

5.3.2 Program-erase endurance/pulse endurance

This process instead of DC sweep a typical AC pulse voltage train as shown in **Figure 19** is applied. During SET pulse the device is bring to ON state or programmed or data is written while during RESET pulse the device is bring to OFF set or data got erased. During read pulse data are read.

Now using similar kind of AC pulse a typical Pulse or Program Erase Endurance characteristics of RRAM device measured by author is (**Figure 20**).

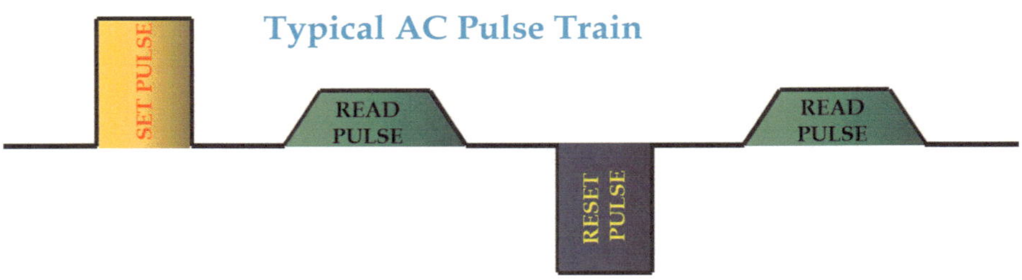

Figure 19.
Typical pulse format used as AC pulse for P/E endurance characteristics.

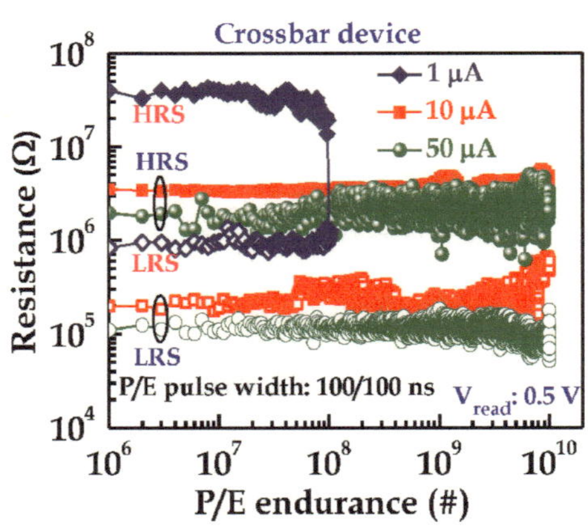

Figure 20.
Typical pulse endurance measured for 3 different program current as 1 μA, 10 μA, and 50 μA by using -2V amplitude and 100 ns pulse width AC pulse for programming and +1 V amplitude and 100 ns pulse width AC pulse for erase [15].

6. Resistive memory potentiality as ultra sensitive biomarker devices

Resistive switching memory (ReRAM) is a type of non-volatile memory that stores information in the resistance state of a solid-state device. It has been primarily explored for its potential applications in data storage and computing. However, researchers have also investigated its use in biosensing applications, including as a biosensor [14–17]. A biosensor is a device that detects and measures biological substances, such as proteins or DNA, by converting a biological response into an electrical signal. Resistive memory works by changing its resistance in response to an applied voltage or current. This change in resistance can be used to detect the presence of biological substances.

Using ReRAM as a biosensor involves leveraging its ability to change resistance in response to certain stimuli, such as the presence of specific molecules or biological substances. This property can be harnessed to detect and analyze various biological targets, including proteins, DNA, and other biomolecules.

In this section we explore elaborately the new potentiality of RRAM devices parallel to storage device as biomarker or biosensor devices. The basic principle behind using ReRAM as a biosensor is to functionalize the surface of the ReRAM device with a specific recognition element, such as antibodies, enzymes, or aptamers. These recognition elements are designed to selectively bind to the target molecules of interest. When the target molecules interact with the recognition elements on the ReRAM surface, it leads to a change in the resistance state of the device. This change in resistance can be measured and correlated with the concentration or presence of the target molecules.

One advantage of using ReRAM as a biosensor is its compatibility with integrated circuit technology. It can be fabricated using standard semiconductor fabrication processes, allowing for the integration of sensing and signal processing functionalities on a single chip. Additionally, ReRAM-based biosensors can offer label-free and real-time detection of biomolecules, making them potentially useful in various applications, such as medical diagnostics, environmental monitoring, and drug discovery.

However, it's important to note that ReRAM-based biosensors are still an active area of research, and several challenges need to be addressed before their widespread adoption. These challenges include improving the sensitivity, selectivity, and stability of the devices, as well as optimizing the functionalization techniques for immobilizing recognition elements on the ReRAM surface.

In summary, ReRAM-based biosensors have the potential to provide label-free and real-time detection of biomolecules. While there is ongoing research in this area, further advancements are needed to fully exploit their capabilities and address existing challenges.

In the following sections I describe about some of the recent research works of using RRAM as biosensor to detect various bioanalytes and pH/H_2O_2 by applying the idea of oxidation-reduction or profanation-deprotonation [14].

The recent advanced research on various potentiality of RRAM has establish the idea that owing to its high sensitivity and specificity, resistive memory biosensors have the potential to revolutionize the way we monitor and diagnose diseases, detect pollutants, and ensure food safety.

One example of a biosensor is the glucose sensor used by diabetics to monitor their blood sugar levels. The sensor contains an enzyme that reacts with glucose in the blood, producing a signal that is converted into a reading of the glucose level. Another example is the biosensor used to detect *E. coli* bacteria in food samples. The biosensor contains antibodies that bind to the bacteria, producing a signal that indicates the presence of the pathogen.

Resistive memory biosensors have been used in various fields, including healthcare and environmental monitoring. In healthcare, they have been used to detect diseases such as cancer and Alzheimer's by detecting specific biomarkers in blood samples. In environmental monitoring, they have been used to detect pollutants in air and water samples. The potential applications of resistive memory biosensors are vast and exciting, with the possibility of detecting a wide range of biological molecules in real-time.

6.1 Advantages of resistive memory as a biosensor

- One of the main advantages of using resistive memory as a biosensor is its high sensitivity. Resistive memory devices can detect very small changes in resistance, which allows them to detect even low concentrations of analytes. This makes them ideal for applications where high sensitivity is required, such as disease diagnosis and environmental monitoring.

- Another advantage of resistive memory biosensors is their low cost and ease of fabrication. Unlike some other types of biosensors, resistive memory devices can be fabricated using standard semiconductor processing techniques, which makes them relatively inexpensive to produce. Additionally, they can be integrated with other electronic components, which makes them easy to use and integrate into existing systems.

6.2 Applications of resistive memory biosensor

6.2.1 Healthcare sector

a. Resistive memory biosensors have the potential to revolutionize healthcare by providing fast, accurate, and non-invasive methods for disease detection and patient monitoring. One example of this is the use of resistive memory biosensors to detect glucose levels in diabetic patients. By measuring changes in resistance caused by glucose binding to the sensor surface, these biosensors can provide real-time monitoring of blood sugar levels without the need for painful finger pricks.

b. Another example is the use of resistive memory biosensors to detect biomarkers in cancer patients. These biosensors can detect minute concentrations of cancer-specific proteins in bodily fluids, allowing for earlier and more accurate diagnosis of the disease. They can also be used to monitor the effectiveness of cancer treatments by tracking changes in biomarker levels over time.

6.2.2 Environmental monitoring

Resistive memory biosensors are becoming increasingly popular in environmental monitoring due to their high sensitivity and specificity. These biosensors can detect pollutants such as heavy metals, pesticides, and organic compounds in air and water with great accuracy. One example of how resistive memory biosensors have been used in environmental monitoring is the detection of nitrogen dioxide (NO_2) in air. By using a resistive memory sensor coated with a thin layer of metal oxide, researchers were able to detect NO_2 levels as low as five parts per billion. This technology has the potential to greatly improve air quality monitoring in urban areas.

6.2.3 Agriculture sector

Resistive memory biosensors can revolutionize agriculture by detecting plant diseases and monitoring soil quality with high accuracy and precision. By measuring the electrical resistance of a conductive filament, these biosensors can detect changes in the chemical composition of soil or plant tissue that indicate the presence of disease or nutrient deficiencies. For example, resistive memory biosensors have been used to detect fungal infections in tomato plants by measuring the levels of specific volatile organic compounds emitted by the plant. This allows farmers to identify and treat the infection before it spreads to other plants. Additionally, resistive memory biosensors can be used to monitor soil moisture levels and nutrient content, providing farmers with real-time data on crop health and growth.

6.2.4 Food safety

Resistive memory biosensors have the potential to revolutionize food safety by detecting pathogens and monitoring food quality. By using resistive memory technology, biosensors can quickly and accurately detect harmful bacteria such as *E. coli* and Salmonella in food products, preventing outbreaks and ensuring consumer safety. In addition to pathogen detection, resistive memory biosensors can also monitor food quality by detecting changes in pH, temperature, and other factors that can affect food spoilage. This allows for early detection of spoilage and can help reduce food waste by ensuring that only safe and high-quality products are sold to consumers.

6.2.5 Security sector

Resistive memory biosensors have the potential to revolutionize security measures by detecting explosives and monitoring air quality in public spaces. These sensors can detect trace amounts of explosive materials, making them an invaluable tool for security personnel. They can also monitor air quality in real-time, alerting authorities to potential hazards before they become a threat. In addition to their use in traditional security settings, resistive memory biosensors can also be used in transportation systems such as airports and subway stations. By monitoring air quality and detecting explosives, these sensors can help prevent terrorist attacks and ensure the safety of travelers.

6.3 Measurement of sensitivity by using RRAM biosensors

In this section first I describe here that how the analytes can be detected in RRAM biosensor. The electrical pulse based sensitivity analysis in RRAM devices can be measured in two methods, one by Capacitance-Voltage measurement (CV) and another is Current-Voltage (IV) measurement. As IV-measurement require less sample volume and perform a fast response, also by this process RRAM filament formation by redox operation also can be more justified so in this section I discuss about the measurement process as well as the reports only based on IV measurement process.

Before the IV measurement first of all a Buffer solution (Tris Buffer) is prepared and then the enzyme and the bioanalyte solution are added to the solution to change the molecular concentration of the solution. Now all the cases there two kinds of solutions are prepared - normal buffer solution with pH 7.4 and the buffer-enzyme-analyte solution with modified pH. Usually the pH enhanced but sometimes it is also reduced depending on the nature of the analytes.

6.3.1 Solution preparation

First a Tris Base is needed to purchase then the TRIS base is dissolved in DI water according to its molecular weight. Then the buffer solution was prepared by adjusting pH value at 7.4 using an HCl solution. Afterward, enzyme followed by bioanalyte are dissolved in the TRIS-HCl (pH 7.4) buffer solution to make a stock solution using the chemical formula. Now addition of bioanalytes will change the concentration of buffer solution. Usually the pH of the analytes mixed buffer solution pH will vary from pH 2 to pH 12 depending on the nature of the analytes is acidic or basic.

6.3.2 IV-measurement

For sensitivity measurement initially the RRAM device is made to be in LRS state by SET the device. Then using micro-pipette put a droplet of buffer solution from stock at the junction of via point or cross-point for RRAM where top and bottom electrode pads merge as shown by device area with arrow in **Figures 12** and **13**.

Now in case of 2D RRAM it is find a better response with either Cu electrode for CBRAM [16, 17] and W [15] or Ir [15] or IrO_x electrode [12–14] for RRAM to act as a Sensing membrane. As Ir or IrO_x is porous in nature and Cu/W is oxidizing metal in nature so the sensing will be better. As in almost majority bioanalyte with buffer solution during application of voltage under the electrolysis operation decomposes and generates H_2O_2 which help to change the oxidation state of sensing membrane electrode and due to oxidation/reduction or protonation/deprotonation mechanism the shotkey barrier of metal/oxide interface changes and the current also changes. Now then with a short duration time a I-V-t sampling operation is undergone which show the shift of current due to change of pH concentration from buffer to bioanalyte mixed buffer solution which gives a value of normalized current and from normalized current the sensitivity can be calculated. **Figure 21(a-f)** below show the typical measurement process of I-Vt sampling of RRAM biosensor for both 2D and 3D resistive memory devices. **Figure 21(a)** describes real time image of the probe-station with the device placed on chuck and probes connected with SMUs dedicated for bias and ground connection during measurement. **Figure 21(b)** shows a typical OM image of a 2×2 μm^2 via-hole memory device , while **Figure 21(c)** illustrates the schematic diagram of RRAM memory structure also used for bio-sensing measurement. **Figure 21(d)** describes the images of the measurement system composed of B1500 and 16440A PGU, and **Figure 21** demonstrates a typical real-time measurement screen shot of long P/E endurance of >10^9 cycles with a small P/E pulse width of 100 ns. Finally, **Figure 21(f)** represents the preparation method of bio-sensing measurement composed of 1μL micropipette and three containers containing stocks of buffer solution, enzyme solution and bioanalyte solution (here creatinine). This setup had already elaborately described in my previous publication [15].

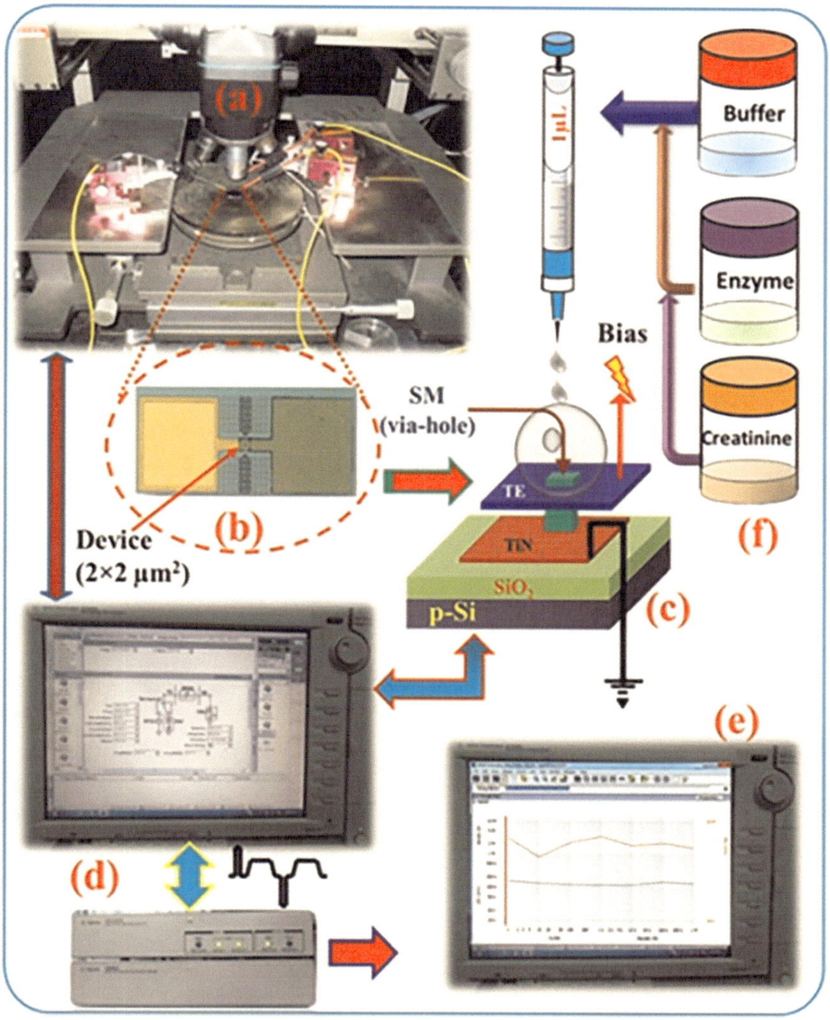

Figure 21.
Experimental setup for the measurement of memory and bio-sensor characteristics. (a) Real time image of the probe-station with the device placed on chuck and probes connected with SMUs dedicated for bias and ground connection during measurement is shown. (b) Typical OEM image of a 2×2 mm² via-hole RRAM memory device. (c) A schematic diagram of RRAM memory structure also used for bio-sensing measurement is described. (d) Images of measurement system composed of B1500 and 16440A PGU are given. (e) A typical real time measurement screen shot of long P/E endurance of >10⁹ cycles with a small P/E pulse width of 100 ns is achieved. (f) A preparation method of biosensing measurement composed of 1 mL micropipette and three containers containing stocks of buffer, buffer and 10 unit creatinine diminase enzyme solution, and 100 nM creatinine solutions are shown [15].

Figure 22 below describes the typical biosensor measurement for crossbar device. Now, bioanalytes reduces first the TE from higher to lower oxidation states ($M^{z+} \rightarrow M^0$). For example, the Ir^{4+} changes to Ir3⁺/Ir^0 oxidation state for Ir or IrO_x TE. At a higher concentration of bioanalyte so i.e.Ir^0 state changes to Ir^{3+}/Ir^{4+} ($Ir^0 \leftrightarrow Ir^{3+} + 3e^-$; $Ir^0 \leftrightarrow Ir^{4+} + 4e^-$). The oxidation-state increment leads to increased work function (Φ_{SB}) of TE metal. This oxidizing current is detected, which decreases with elevated bioanalytes concentration. Therefore, the Φ_{SB} value at the TE/SM interface increases gradually as well as the sensing current decreases with increasing bioanalytes concentration.

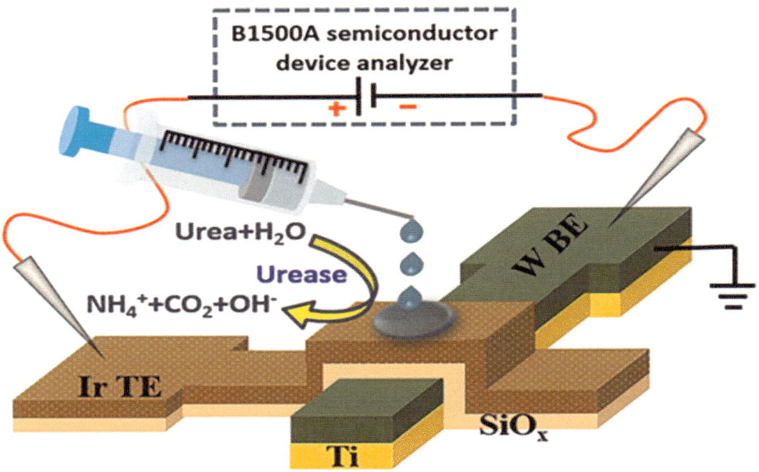

Figure 22.
A 3D illustration of typical experimental process of urea sensing through our fabricated cross-point structure which is connected to B1500 measurement unit to perform the electrical characteristics.

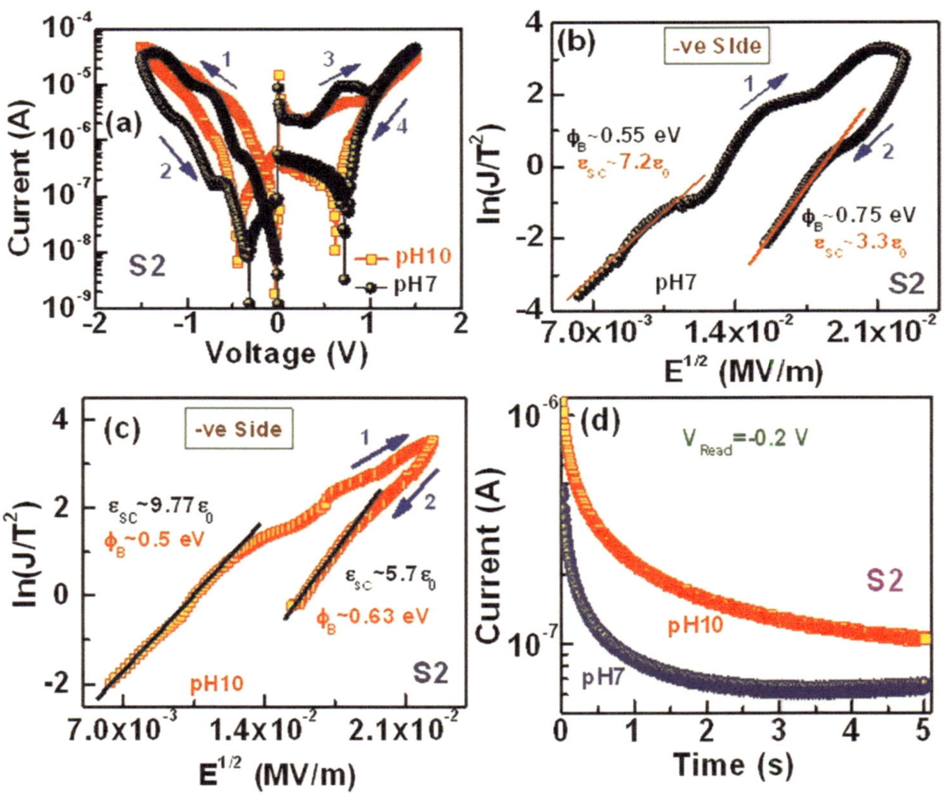

Figure 23.
(a) Typical IV characteristics with pH7 (buffer) and pH10 (some bioanalyte mixed buffer solution) of RRAM biosensor (b) & (c) typical shotkey barrier height under positive and negative bias, respectively, and (d) IV-t sampling curve measured by keeping electrolyte solution with pH7 and pH10 at sensing membrane for 5s and measuring IV characteristics.

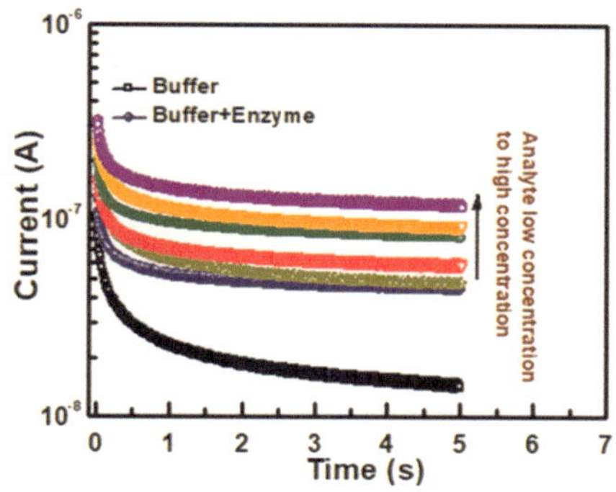

Figure 24.
IV-t sampling curve shift from buffer solution reference with varying concentration of bioanalyte solution.

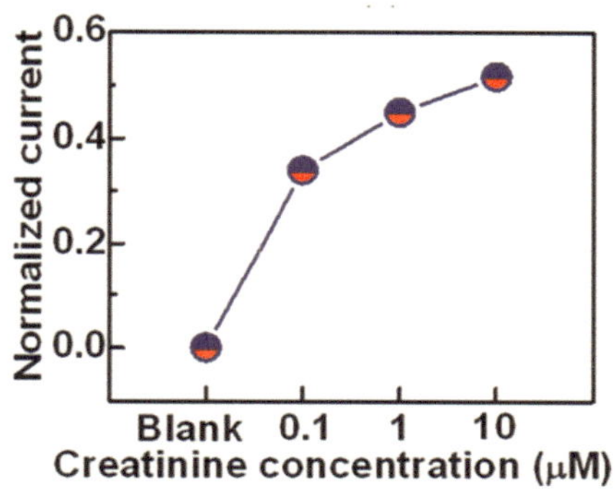

Figure 25.
A typical normalized current plot with varying creatinine concentration as reported in [15].

During this IV-t sampling the typical IV and change in shotkey barrier height and the oxidizing current changes due to pH variation from different concentration of bioanalytes with reference to buffer solution are shown in **Figure 23**.

Now normalized current can be calculated by using the following formula

$$\textit{normalized current} = \frac{I_c - I_0}{I_0} \quad \begin{array}{l} \mathbf{I_0} = \textit{Current change for buffer} + \textit{enzyme} \\ \mathbf{I_c} = \textit{Current change for bioanalytes} \end{array} \tag{1}$$

A typical normalized current plot can be drawn from the typical IV-t sampling curve as shown in **Figure 24** by measuring the changes of final current bioanalyte with varying concentration and the typical curve will be looking as **Figure 25**.

As most of the bioanalytes after electrolysis decomposed to generate H_2O_2, finally sensitivity can be calculated by the relation between H_2O_2 concentration (α) and normalized current shift (ΔI_N) which is purely exponential which can be obtained by linearly fitted **Figure 23**. And the obtained relation is

$$\alpha = \exp\left[\frac{\Delta I}{a} - \frac{b}{a}\right] \tag{2}$$

where a is the intercept and b is the slope of linearly fitted curve at shown in **Figure 25**.

Now there are multiple reports on various bioanalyte sensing [12–17, 76–78].

Table 6 shows the different RRAM structures with different analytes detection with lowest concentration.

Structure	Sensing Membrane	Bioanalyte	Detection Method	Lower Detection Level	Usefulness
IrOx/GdOx/W -CP RRAM [12]	IrO_x	Urea	CV	1 mM	High level of Urea can cause Ulcers, Indigestion and several Renal Diseases, so early detection of Urea is required
IrOx/Al_2O_3/W -CP RRAM [13]	IrO_x	Glucose	CV	10pM	Glucose detection is an important to prevent hyperglucemia and hypoglucemia
W/GeO_x/W-2D RRAM [14]	W	Sarcosine	CV	50 pM	Sarcosine is indicator of Prostrate Cancer so low concentration detection can help for early prevention and treatment.
Ir/SiO_x/TiN -2 [15]	Ir	Creatinine	IV-t sampling	100 nM	High level of Creatinine can cause High blood pressure, diabetes and Kidney failure and several other Renal Diseases, so early detection of Creatinine can help for early treatment.
Cu/AlOx/a-COx/ TiNxOy/TiN-2D CBRAM [16]	Cu	Saliva	IV-t sampling	1 pM	Non invasive way Glucose detection from Saliva by using lower sample volume.
Cu/Ir/TiNxOy/TiN-2D CBRAM [17]	Cu & Ir	Tributyrin	IV-t sampling	1 pM	High concentration of Tributyrin may cause gastrointestinal disturbances, such as diarrhea, bloating, and stomach cramps etc. So early detection can prevent these.
IrOx/Al_2O_3/TaOx/ MoS_2/TiN-2D RRAM [76]	IrOx	Ascorbic Acid	IV-t sampling	1 pM	High concentration of Ascorbic Acid might cause side effects such as stomach cramps, nausea, heartburn, and headache. So early detection can prevent these.
Cu/MoS_2/TiN-2D CBRAM [77]	Cu	Dopamine	IV-t sampling	10 pM	Nausea, vomiting, orthostatic hypotension, headache, dizziness, and cardiac arrhythmia are the most common side effect of dopamine, so early detection can prevent these.
W/Al_2O_3/TaOx/ TiN-2D RRAM [78]	W	Dopamine	IV-t sampling	1 pM	Nausea, vomiting, orthostatic hypotension, headache, dizziness, and cardiac arrhythmia are the most common side effect of dopamine, so early detection can prevent these.

Table 6.
Resistive memory-based biosensors with different analyte detection with lower detected concentration.

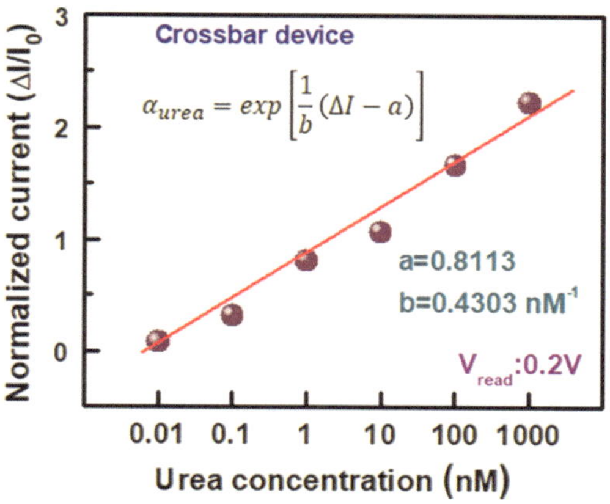

Figure 26.
A typical normalized current plot with varying urea concentration by linearly fitting the curve as per shown in ***Figure 25*** *to calculate the sensitivity.*

Now based on the reports below show some of the equations of bioanalyte decomposition like urea, creatinine, Sarcosine and LOXL2. In these Creatinine and urea are very important for kidney disease identification where as Sarcosine and LOXL2 are important for prostrate and breast cancer detection.

$$N - Lysine + O_2 + H_2O \overset{LOXL2}{\leftrightarrow} Allysine + NH_3 + H_2O_2 \quad (3)$$

$$Sarcosine + O_2 + H_2O \overset{SO_x}{\leftrightarrow} Formaldehyde + Glycin + H_2O_2 \quad (4)$$

$$Urea + H_2O \overset{Urease}{\leftrightarrow} 2NH_4^+ + CO_2 + 2OH^- \quad (5)$$

$$creatinime + H_2O \overset{\substack{creatimine \\ deaminase}}{\leftrightarrow} N - methylthydantion + NH_4^+ + OH^- \quad (6)$$

The typical sensitivity plot of urea sensing by using the formula in (2) is shown below (**Figure 26**).

The lowest sensitivity will be denoted by the corresponding bioanalyte concentration values for which the normalized current of the bioanalyte will be minimum from the curve.

6.4 Challenges of resistive memory as a biosensor

One of the main challenges of using resistive memory as a biosensor is achieving high sensitivity and selectivity. While resistive memory biosensors have shown promising results in detecting biomolecules, there is still a need to improve their performance in terms of accuracy and reliability. Researchers are currently exploring various approaches to enhance the sensing capabilities of resistive memory biosensors, such as incorporating functionalized nanomaterials or optimizing the device architecture.

Another challenge is the integration of resistive memory biosensors into practical applications. The fabrication process of resistive memory devices can be complex and costly, which limits their scalability and commercialization. To address this issue, researchers are investigating new materials and manufacturing techniques that can simplify the production process and reduce the cost of resistive memory biosensors.

6.5 Future of resistive memory as a biosensor

The future of resistive memory biosensors is exciting and full of potential. As technology advances, we can expect to see these biosensors being used in a variety of fields, including agriculture, food safety, and security. In agriculture, resistive memory biosensors can be used to detect plant diseases and monitor soil quality. This can lead to more efficient and sustainable farming practices, ultimately increasing crop yields and reducing the use of harmful pesticides.

In the field of food safety, resistive memory biosensors can be used to detect pathogens and monitor food quality. This can help prevent food borne illnesses and ensure that consumers are getting safe and high-quality products. In terms of security, resistive memory biosensors have the potential to be used for detecting explosives and monitoring air quality in public spaces. This can help keep people safe and prevent dangerous situations from occurring.

Overall, the future of resistive memory biosensors is bright and full of possibilities. With continued research and development, we can expect to see these biosensors being used in even more innovative ways in the years to come.

7. Conclusion

In conclusion, resistive memory can be found to have a completely new are of application than its conventional high density memory utilization is acting as biosensors. RRAM biosensors have the potential to revolutionize various fields, including healthcare, environmental monitoring, agriculture, food safety, security, energy, and biomedical research. By using conductive filaments to detect changes in electrical resistance, these biosensors offer high sensitivity, specificity, and selectivity compared to other types of biosensors. They also have the advantage of being low-cost, label-free, and easy to fabricate.

However, there are still challenges that need to be addressed, such as improving the stability and reproducibility of the conductive filaments, optimizing the sensing performance, and reducing the interference from external factors. Nevertheless, with ongoing research and development, resistive memory biosensors hold great promise for enhancing our ability to detect and monitor various biological and environmental parameters.

Author details

Sourav Roy
Department of Electronics and Communication Engineering, CHRIST University, Kengeri Campus, Karnataka, Bangalore, India

*Address all correspondence to: sourav.roy@christuniversity.in

References

[1] Waser R. Nanotechnology. KGaA, Germany: John Wiley & Sons; 2008

[2] Lacaze PC, Lacroix JC. Non-volatile Memories. London, UK: John Wiley & Sons; 2014

[3] The Rise of the Flash Memory Market: Its Impact on Firm Behavior and Global Semiconductor Trade Patterns | United States International Trade Commission. (n.d.). Available from: https://www.usitc.gov/staff_publications/jice/rise_flash_memory_market_its_impact_firm_behavior

[4] Keeney S. A 130 nm generation high density Etox/sup TM/flash memory technology. In: International Electron Devices Meeting. Technical Digest (Cat. No.01CH37224). Washington, D.C., USA: IEEE International Electron Devices Meeting. 2001. DOI: 10.1109/iedm.2001.979398

[5] Kubota T, Ando K, Muramatsu S. FLASH memory data retention reliability and the floating gate/tunnel SiO_2 interface characteristics. Applied Surface Science. 1997;**117–118**:253-258. DOI: 10.1016/s0169-4332(97)80090-6

[6] Emerging Non Volatile Memory (NVM) Technology & Market Trends: 2014 Report by Yole Developpement. (n.d.). PPT. Available from: https://www.slideshare.net/Yole_Developpement/yole-emerging-nonvolatilememorysample

[7] Kozicki M, Gopalan C, Balakrishnan M, Park M, Mitkova M. Nonvolatile memory based on solid electrolytes. In: Proceedings, 2004 IEEE Computational Systems Bioinformatics Conference, Stanford, CA, USA. 2004. DOI: 10.1109/nvmt.2004.1380792

[8] Waser R, Dittmann R, Staikov G, Szot K. Redox-based resistive switching memories—Nanoionic mechanisms, prospects, and challenges. Advanced Materials. 2009;**21**(25–26):2632-2663. DOI: 10.1002/adma.200900375

[9] Valov I, Waser R, Jameson JR, Kozicki MN. Electrochemical metallization memories—Fundamentals, applications, prospects. Nanotechnology. 2011;**22**(25):254003. DOI: 10.1088/0957-4484/22/25/254003

[10] Prakash A, Jana D, Maikap S. TaO x -based resistive switching memories: Prospective and challenges. Nanoscale Research Letters. 2013;**8**(1):1-17. DOI: 10.1186/1556-276x-8-418

[11] Roy S, Niu G, Wang Q, Wang Y, Zhang Y, Wu H, et al. Toward a reliable synaptic simulation using Al-doped HfO_2 RRAM. ACS Applied Materials & Interfaces. February 11 2020;**12**(9): 10648-10656. DOI: 10.1021/acsami.9b21530

[12] Kumar P, Maikap S, Ginnaram S, Qiu JT, Jana D, Chakrabarti S, et al. Cross-point resistive switching memory and urea sensing by using annealed GdOxFilm in IrOx/GdOx/W structure for biomedical applications. Journal of the Electrochemical Society. 2017; **164**(4):B127-B135. DOI: 10.1149/2.1011704jes

[13] Chakrabarti S, Maikap S, Samanta S, Jana S, Roy A, Qiu JT. Scalable cross-point resistive switching memory and mechanism through an understanding of H_2O_2/glucose sensing using an IrOx/Al_2O_3/W structure. Physical Chemistry Chemical Physics. 2017;**19**(38): 25938-25948. DOI: 10.1039/c7cp05089e

[14] Samanta S, Rahaman SZ, Roy A, Jana S, Chakrabarti S, Panja R, et al. Understanding of multi-level resistive switching mechanism in GeOx through

redox reaction in H_2O_2/sarcosine prostate cancer biomarker detection. Scientific Reports. September 11 2017; 7(1):11240. DOI: 10.1038/s41598-017-11657-4

[15] Roy S, Roy A, Panja R, Samanta S, Chakrabarti S, Yu PL, et al. Comparison of resistive switching characteristics by using e-gun/sputter deposited SiOx film in W/SiOx/TiN structure and pH/creatinine sensing through iridium electrode. Journal of Alloys and Compounds. 2017;**726**:30-40. DOI: 10.1016/j.jallcom.2017.07.304

[16] Ginnaram S, Qiu JT, Maikap S. Controlling Cu migration on resistive switching, artificial synapse, and glucose/saliva detection by using an optimized AlOx interfacial layer in a-COx-based conductive bridge random access memory. ACS Omega. 2020;**5**(12): 7032-7043. DOI: 10.1021/acsomega.0c00795

[17] Dutta M, Maikap S, Qiu JT. Controlling conductive filament and tributyrin sensing using an optimized porous iridium interfacial layer in Cu/Ir/TiN x O y/TiN. Advanced Electronic Materials. 2018;**5**:1800288. DOI: 10.1002/aelm.201800288

[18] Yang JJ, Zhang MX, Strachan JP, Miao F, Pickett MD, Kelley RD, et al. High switching endurance in TaOx memristive devices. Applied Physics Letters. 2010;**97**(23):232102. DOI: 10.1063/1.3524521

[19] Belmonte A, Witters T, Covello A, Vereecke G, Franquet A, Spampinato V, et al. Co active electrode enhances CBRAM performance and scaling potential. In: 2019 IEEE International Electron Devices Meeting (IEDM), San Francisco, CA, USA. 2019. DOI: 10.1109/iedm19573.2019.8993527

[20] Rahaman SZ, Maikap S, Das A, Prakash A, Wu YH, Lai CS, et al. Enhanced nanoscale resistive switching memory characteristics and switching mechanism using high-Ge-content Ge0.5Se0.5 solid electrolyte. Nanoscale Research Letters. 2012;**7**(1):614. DOI: 10.1186/1556-276x-7-614

[21] Jana D, Samanta S, Roy S, Lin YF, Maikap S. Observation of resistive switching memory by reducing device size in a new Cr/CrOx/TiOx/TiN structure. Nano-Micro Letters. 2015;**7**(4):392-399. DOI: 10.1007/s40820-015-0055-3

[22] Lee MJ, Lee CB, Lee D, Lee SR, Chang M, Hur JH, et al. A fast, high-endurance and scalable non-volatile memory device made from asymmetric Ta_2O_{5-x}/TaO_{2-x} bilayer structures. Nature Materials. 2011;**10**(8):625-630. DOI: 10.1038/nmat3070

[23] Hickmott TW. Low-frequency negative resistance in thin anodic oxide films. Journal of Applied Physics. 1962; **33**(9):2669-2682. DOI: 10.1063/1.1702530

[24] Waser R, Aono M. Nanoionics-based resistive switching memories. Nature Materials. 2007;**6**(11):833-840. DOI: 10.1038/nmat2023

[25] Sawa A. Resistive switching in transition metal oxides. Materials Today. 2008;**11**(6):28-36. DOI: 10.1016/s1369-7021(08)70119-6

[26] Chen YS, Lee HY, Chen PS, Wu TY, Wang CC, Tzeng PJ, et al. An ultrathin forming-free HfO_x resistance memory with excellent electrical performance. IEEE Electron Device Letters. 2010; **31**(12):1473-1475. DOI: 10.1109/led.2010.2081658

[27] Banerjee W, Rahaman SZ, Maikap S. Excellent uniformity and multilevel

operation in formation-free low power resistive switching memory using IrOx/AlOx/W cross-point. Japanese Journal of Applied Physics. 2012;**51**(4S):04DD10. DOI: 10.1143/jjap.51.04dd10

[28] Jeong HY, Lee JY, Ryu MK, Choi SY. Bipolar resistive switching in amorphous titanium oxide thin film. Physica Status Solidi (RRL) - Rapid Research Letters. February 2010;**4**(1–2):28-30. DOI: 10.1002/pssr.200903383

[29] Xu N, Gao B, Liu L, Sun B, Liu X, Han R, et al. A unified physical model of switching behavior in oxide-based RRAM. In: 2008 Symposium on VLSI Technology, Kyoto, Japan. 2008. DOI: 10.1109/vlsit.2008.4588578

[30] Lin CY, Wu CY, Wu CY, Tseng TY, Hu C. Modified resistive switching behavior of ZrO_2 memory films based on the interface layer formed by using Ti top electrode. Journal of Applied Physics. 2007;**102**(9):094101. DOI: 10.1063/1.2802990

[31] Muenstermann R, Menke T, Dittmann R, Waser R. Coexistence of filamentary and homogeneous resistive switching in Fe-doped $SrTiO_3$ thin-film Memristive devices. Advanced Materials. 2010;**22**(43):4819-4822. DOI: 10.1002/adma.201001872

[32] Lee HY, Chen YS, Chen PS, Gu PY, Hsu YY, Wang SM, et al. Evidence and solution of over-RESET problem for HfO_X based resistive memory with sub-ns switching speed and high endurance. In: 2010 International Electron Devices Meeting, San Francisco, CA, USA. 2010. DOI: 10.1109/iedm.2010.5703395

[33] Cao X, Li X, Gao X, Yu W, Liu X, Zhang Y, et al. Forming-free colossal resistive switching effect in rare-earth-oxide Gd_2O_3 films for memristor applications. Journal of Applied Physics. October 1 2009;**106**(7):073723. DOI: 10.1063/1.3236573

[34] Robertson J. High dielectric constant oxides. European Physical Journal Applied Physics. 2004;**28**:265-291. DOI: 10.1051/epjap:2004206

[35] Birks N, Meier GH, Pettit FS. Introduction to the High Temperature Oxidation of Metals. Cambridge, UK: Cambridge University Press; 2006

[36] Dean JA, Lange NA. Lange's Handbook of Chemistry. Laramie, Wyoming, USA: McGraw-Hill; 1999

[37] Wei Z, Kanzawa Y, Arita K, Katoh Y, Kawai K, Muraoka S, et al. Reliable TaOx ReRAM and direct evidence of redox reaction mechanism. In: Highly 2008 IEEE International Electron Devices Meeting, San Francisco, CA, USA. 2008. DOI: 10.1109/iedm.2008.4796676

[38] Lee AR, Baek GH, Kim TY, Ko WB, Yang SM, Kim J, et al. Memory window engineering of Ta_2O_{5-x} oxide-based resistive switches via incorporation of various insulating frames. Scientific Reports. 2016;**6**(1):1-9. DOI: 10.1038/srep30333

[39] Chen YY, Goux L, Clima S, Govoreanu B, Degraeve R, Kar GS, et al. Endurance/retention trade-off on HfO_2 metal cap 1T1R bipolar RRAM. IEEE Transactions on Electron Devices. 2013;**60**(3):1114-1121. DOI: 10.1109/ted.2013.2241064

[40] Zhou Y, Yew KS, Ang DS, Kawashima T, Bera MK, Zhang HZ, et al. White-light-induced disruption of nanoscale conducting filament in hafnia. Applied Physics Letters. 2015;**107**(7):1-5. DOI: 10.1063/1.4929324

[41] Joshua Yang J, Miao F, Pickett MD, Ohlberg DAA, Stewart DR, Lau CN,

et al. The mechanism of electroforming of metal oxide memristive switches. Nanotechnology. 2009;**20**(21):215201. DOI: 10.1088/0957-4484/20/21/215201

[42] Jana D, Dutta M, Samanta S, Maikap S. RRAM characteristics using a new Cr/GdOx/TiN structure. Nanoscale Research Letters. December 2014;**9**(1):1-9. DOI: 10.1186/1556-276x-9-680

[43] Chien W, Chen Y, Chen Y, Chuang AT, Lee F, Lin Y, et al. A forming-free WOx resistive memory using a novel self-aligned field enhancement feature with excellent reliability and scalability. In: 2010 International Electron Devices Meeting, San Francisco, CA, USA. 2010. DOI: 10.1109/iedm.2010.5703390

[44] Chakrabarti S, Ginnaram S, Jana S, Wu ZY, Singh K, Roy A, et al. Negative voltage modulated multi-level resistive switching by using a Cr/BaTiOx/TiN structure and quantum conductance through evidence of H_2O_2 sensing mechanism. Scientific Reports. 2017; **7**(1):1-13. DOI: 10.1038/s41598-017-05059-9

[45] Samanta S, Gong X, Zhang P, Han K, Fong X. Bipolar resistive switching and synaptic characteristics modulation at sub-μA current level using novel Ni/SiOx/W cross-point structure. Journal of Alloys and Compounds. 2019;**805**:915-923. DOI: 10.1016/j.jallcom.2019.07.050

[46] Gao Q, Huang A, Hu Q, Zhang X, Chi Y, Li R, et al. Stability and repeatability of a karst-like hierarchical porous silicon oxide-based memristor. ACS Applied Materials & Interfaces. 2019;**11**(24):21734-21740. DOI: 10.1021/acsami.9b06855

[47] Belmonte A, Kim W, Chan BT, Heylen N, Fantini A, Houssa M, et al. A thermally stable and high-performance 90-nm Al_2O_3 Cu-based 1T1R CBRAM cell. IEEE Transactions on Electron Devices. 2013;**60**(11):3690-3695. DOI: 10.1109/ted.2013.2282000

[48] Mehonic A, Cueff S, Wojdak M, Hudziak S, Jambois O, Labbé C, et al. Resistive switching in silicon suboxide films. Journal of Applied Physics. 2012; **111**(7):1-9. DOI: 10.1063/1.3701581

[49] Jana D, Roy S, Panja R, Dutta M, Rahaman SZ, Mahapatra R, et al. Conductive-bridging random access memory: Challenges and opportunity for 3D architecture. Nanoscale Research Letters. 2015;**10**(1):1-23. DOI: 10.1186/s11671-015-0880-9

[50] Kozicki M, Park M, Mitkova M. Nanoscale memory elements based on solid-state electrolytes. IEEE Transactions on Nanotechnology. 2005; **4**(3):331-338. DOI: 10.1109/tnano.2005.846936

[51] Vianello E, Molas G, Longnos F, Blaise P, Souchier E, Cagli C, et al. Sb-doped GeS_2 as performance and reliability booster in conductive bridge RAM. In: 2012 International Electron Devices Meeting, San Francisco, CA, USA. 2012. DOI: 10.1109/iedm.2012.6479145

[52] Lee D, Oukassi S, Molas G, Carabasse C, Salot R, Perniola L. Memory and energy storage dual operation in chalcogenide-based CBRAM. IEEE Journal of the Electron Devices Society. 2017;**5**(4):283-287. DOI: 10.1109/jeds.2017.2693220

[53] Sakamoto T, Banno N, Iguchi N, Kawaura H, Sunamura H, Fujieda S, et al. A Ta_2O_5 solid-electrolyte switch with improved reliability. In: 2007 IEEE Symposium on VLSI Technology, Kyoto,

Japan. 2007. DOI: 10.1109/vlsit.2007.4339718

[54] Rahaman SZ, Maikap S, Tien TC, Lee HY, Chen WS, Chen FT, et al. Excellent resistive memory characteristics and switching mechanism using a Ti nanolayer at the Cu/TaOx interface. Nanoscale Research Letters. 2012;7(1):1-11. DOI: 10.1186/1556-276x-7-345

[55] Xiao B, Gu T, Tada T, Watanabe S. Conduction paths in Cu/amorphous-Ta_2O_5/Pt atomic switch: First-principles studies. Journal of Applied Physics. 2014; **115**(3). DOI: 10.1063/1.4861724

[56] Roy S, Maikap S, Sreekanth G, Dutta M, Jana D, Chen Y, et al. Improved resistive switching phenomena and mechanism using Cu-Al alloy in a new Cu: AlOx/TaOx/TiN structure. Journal of Alloys and Compounds. 2015;**637**: 517-523. DOI: 10.1016/j.jallcom.2015.02.168

[57] Xiao B, Watanabe S. Interface structure in Cu/Ta_2O_5/Pt resistance switch: A first-principles study. ACS Applied Materials & Interfaces. 2015; 7(1):519-525. DOI: 10.1021/am5066656

[58] Panja R, Roy S, Jana D, Maikap S. Impact of device size and thickness of Al_2O_3 film on the Cu pillar and resistive switching characteristics for 3D cross-point memory application. Nanoscale Research Letters. 2014;**9**(1):1-11. DOI: 10.1186/1556-276x-9-692

[59] Woo J, Belmonte A, Redolfi A, Hwang H, Jurczak M, Goux L. Introduction of WO3 layer in a Cu-based Al_2O_3 conductive bridge RAM system for robust cycling and large memory window. IEEE Journal of the Electron Devices Society. 2016;**4**(3):163-166. DOI: 10.1109/jeds.2016.2526632

[60] Maikap S, Panja R, Jana D. Copper pillar and memory characteristics using Al_2O_3 switching material for 3D architecture. Nanoscale Research Letters. 2014;**9**(1):1-9. DOI: 10.1186/1556-276x-9-366

[61] Kim HJ, Park TH, Yoon KJ, Seong WM, Jeon JW, Kwon YJ, et al. Fabrication of a Cu-cone-shaped cation source inserted conductive bridge random access memory and its improved switching reliability. Advanced Functional Materials. 2019;**29**(8):1-13. DOI: 10.1002/adfm.201806278

[62] Schindler C, Thermadam SCP, Waser R, Kozicki MN. Bipolar and unipolar resistive switching in Cu-doped SiO_2. IEEE Transactions on Electron Devices. 2007;**54**(10):2762-2768. DOI: 10.1109/ted.2007.904402

[63] Tappertzhofen S, Menzel S, Valov I, Waser R. Redox processes in silicon dioxide thin films using copper microelectrodes. Applied Physics Letters. 2011;**99**(20):1-3. DOI: 10.1063/1.3662013

[64] Chen W, Fang R, Balaban MB, Yu W, Gonzalez-Velo Y, Barnaby HJ, et al. A CMOS-compatible electronic synapse device based on Cu/SiO_2/W programmable metallization cells. Nanotechnology. 2016;**27**(25):255202. DOI: 10.1088/0957-4484/27/25/255202

[65] Wu F, Si S, Cao P, Wei W, Zhao X, Shi T, et al. Interface engineering via MoS_2 insertion layer for improving resistive switching of conductive-bridging random access memory. Advanced Electronic Materials. 2019; **5**(4):1800747. DOI: 10.1002/aelm.201800747

[66] Ginnaram S, Qiu JT, Maikap S. Role of the Hf/Si interfacial layer on the high

performance of MoS_2-based conductive bridge RAM for artificial synapse application. IEEE Electron Device Letters. 2020;**41**(5):709-712. DOI: 10.1109/led.2020.2980625

[67] Hong X, Loy DJ, Dananjaya PA, Tan F, Ng C, Lew W. Oxide-based RRAM materials for neuromorphic computing. Journal of Materials Science. 2018;**53**(12):8720-8746. DOI: 10.1007/s10853-018-2134-6

[68] Seok JY, Song SJ, Yoon JH, Yoon KJ, Park TH, Kwon DE, et al. A review of three-dimensional resistive switching cross-bar Array memories from the integration and materials property points of view. Advanced Functional Materials. 2014;**24**(34):5316-5339. DOI: 10.1002/adfm.201303520

[69] Hu Q, Park MR, Abbas H, Kang TS, Yoon TS, Kang CJ. Forming-free resistive switching characteristics in tantalum oxide and manganese oxide based crossbar array structure. Microelectronic Engineering. 2018; **190**:7-10. DOI: 10.1016/j.mee.2018.01.005

[70] Maikap S, Senapati A. Ru conducting filament based cross-point resistive switching memory for future low power operation. In: 2020 IEEE Silicon Nanoelectronics Workshop (SNW), Honolulu, HI, USA. 2020. DOI: 10.1109/snw50361.2020.9131659

[71] Wu H, Li X, Wu M, Huang F, Yu Z, Qian H. Resistive switching performance improvement of Ta_2O_{5-x} TaO_y bilayer ReRAM devices by inserting AlO_δ barrier layer. IEEE Electron Device Letters. 2014;**35**(1):39-41. DOI: 10.1109/led.2013.2288311

[72] Yu S, Chen HY, Gao B, Kang J, Wong HSP. HfOx-based vertical resistive switching random access memory suitable for bit-cost-effective three-dimensional cross-point architecture. ACS Nano. 2013;**7**(3):2320-2325. DOI: 10.1021/nn305510u

[73] Persson K, Ram MS, Kilpi O, Borg M, Wernersson L. Cross-point arrays with low-power ITO-HfO 2 resistive memory cells integrated on vertical III-V nanowires. Advanced Electronic Materials. 2020;**6**(6):2000154. DOI: 10.1002/aelm.202000154

[74] Lee F, Lin Y, Lee M, Chien W, Lung H, Hsieh K, et al. A novel cross point one-resistor (0T1R) conductive bridge random access memory (CBRAM) with ultra low set/reset operation current. In: 2012 IEEE Symposium on VLSI Technology (VLSIT), Honolulu, HI, USA. 2012. DOI: 10.1109/vlsit.2012.6242464

[75] Kim S, Park J, Kim TH, Hong K, Hwang Y, Park BG, et al. 4-bit multilevel operation in overshoot suppressed Al_2O_3/TiOx resistive random-access memory crossbar array. Advanced Intelligent Systems. 2022;**4**(9):2100273. DOI: 10.1002/aisy.202100273

[76] Samanta S, Maikap S, Roy A, Jana S, Qiu JT. Effects of W/Ir top electrode on resistive switching and dopamine sensing by using optimized TaOx-based memory platform. Advanced Materials Interfaces. 2017;**4**(24):1700959. DOI: 10.1002/admi.201700959

[77] Dutta M, Ginnaram S, Roy A, Maikap S. Cu filament based resistive switching and oxidation reduction through dopamine sensing in novel Cu/MoS2/TinN structure. In: 2018 IEEE International Memory Workshop (IMW), Kyoto, Japan. 2018. DOI: 10.1109/imw.2018.8388825

[78] Qiu JT, Samanta S, Dutta M, Ginnaram S, Maikap S. Controlling resistive switching by using an optimized MoS_2 interfacial layer and the role of top electrodes on ascorbic acid sensing in TaOx-based RRAM. Langmuir. 2019;**35**(11):3897-3906. DOI: 10.1021/acs.langmuir.8b04090

Chapter 8

Non-Idealities in Memristor Devices and Methods of Mitigating Them

Muhammad Ahsan Kaleem, Jack Cai, Yao-Feng Chang, Roman Genov and Amirali Amirsoleimani

Abstract

One of the main issues that memristors face, like other hardware components, is non-idealities (that can arise from long-term usage, low-quality hardware, etc.). In this chapter, we discuss some ways of mitigating the effects of such non-idealities. We consider both hardware-based solutions and universal solutions that do not depend on hardware or specific types of non-idealities, specifically in the context of memristive neural networks. We compare such solutions both theoretically and empirically using simulations. We also explore the different non-idealities in depth, such as device faults, endurance, retention, and finite conductance states, considering what causes them and how they can be avoided, and present ways of simulating these non-idealities in software.

Keywords: memristors, memristor non-idealities, non-idealities simulation, deep neural networks, ensemble methods

1. Introduction

Memristors are electronic components that are able to speed up vector-matrix multiplication (VMM) operations by performing them entirely in memory and thus have many applications such as accelerating machine learning. **Figure 1a** depicts the mapping of memristive VMM to artificial neural networks. One of the main problems faced by memristors that hinder their more mainstream use is non-idealities (**Figure 1b**). These are phenomena that negatively affect their performance, and thus, it is important to find ways to mitigate their effects. We begin this chapter by briefly describing some of the non-idealities commonly present in memristors, followed by a discussion on simulations of memristor devices and circuits with non-idealities, and we conclude by describing methods of mitigating the effects of non-idealities in memristors.

IntechOpen

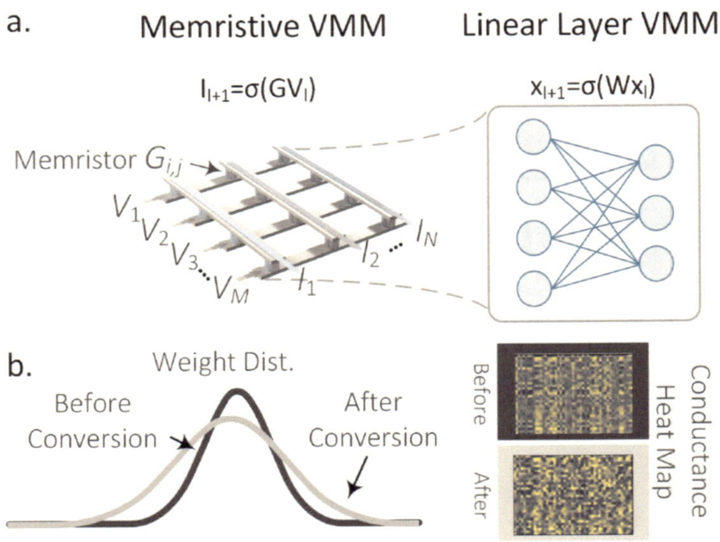

Figure 1.
(a) Memristive vector matrix multiplication (VMM) and its topological mapping to artificial neural networks. (b) Weight distribution shift introduced by memristor non-idealities.

2. Non-idealities in memristors

2.1 Device faults

Memristors have low and high resistance states (LRS and HRS), so device faults are non-idealities where individual memristors become stuck in one of these states. A memristor stuck in the HRS is effectively an open circuit, while a memristor stuck in the LRS is effectively a short circuit.

2.2 Endurance

Endurance is the non-ideality that essentially refers to the phenomenon of wear and tear; like any other hardware components, the performance of memristors degrades over time with use (SET and RESET cycles).

2.3 Random telegraph noise

Random telegraph noise refers to the phenomenon of sudden, random fluctuations in the properties of electronic devices. In the context of memristors, this random fluctuation can affect the resistance, thus resulting in a change in the conductance state of the memristor and causing non-ideal behavior to be exhibited.

2.4 Retention

Memristors have a finite retention time, meaning that they eventually begin to lose the information that they have stored over time. Retention refers to the non-ideal behavior that arises due to this phenomenon.

DOI: http://dx.doi.org/10.5772/intechopen.1003837

2.5 Finite conductance states

Ideally, memristors are assumed to have an unlimited amount of conductance states. However, in reality there can only be a finite amount of conductance states. This causes non-ideal behavior because the memristor cannot be set to the exact conductance value that it is supposed to be set to and instead has to be rounded to the nearest conductance value that is supported by the memristor.

2.6 Device-to-device and cycle-to-cycle variability

Overall, different devices exhibit various levels of non-idealities. These non-idealities can be viewed holistically as device-to-device (D2D) variability or cycle-to-cycle (C2C) variability. The coefficient of variation (C_v) is defined as the ratio of the standard deviation of the SET/RESET voltages with the means, with typical values ranging from less than 1% to up to around 6% for C2C SET/RESET variability, and up to 35% reported for D2D SET/RESET variability [1–4]. For example, Kumar et al. [4] reported 0.2%/1.07% and 2.64%/10.13% for C2C and D2D SET/RESET variability for $Al/Y_2O_3/GZO$ memristor on 30×25 size array.

3. Simulation of memristor devices and circuits with non-idealities

3.1 Overview

While enabling exciting possibilities in various fields, memristors exhibit non-ideal behaviors that can significantly impact their performance and reliability in practical applications. Therefore, it is crucial to accurately model and simulate memristor devices and circuits, taking these non-idealities into account. In this section, we delve into the complexities of simulating memristor behavior, with a particular focus on addressing non-idealities. Starting with a single memristor device, we can explore the intricacies of its operation and the influence of non-idealities on its behavior. Subsequently, we can scale up our simulations to encompass entire memristor-based systems, providing insights into how non-idealities propagate through interconnected components and affect the overall performance of complex applications.

3.2 Simulation of memristor device

Simulating memristor devices is a fundamental aspect of memristor research. To achieve accurate simulations, one must consider various non-ideal behaviors that can arise in memristor devices. These non-idealities encompass factors such as variability in resistance states, endurance issues, and parasitic effects. Mathematical models, which include non-linear differential equations and piecewise linear approximations, are commonly used to describe memristor behavior. In 1971, Chua predicted the flux (current) controlled memristor and the charge (voltage) controlled memristor in his work "Memristor – The Missing Circuit Element" [5], which is described by the following relationship:

$$i(t) = W(\phi(t))v(t), W(q) \equiv \frac{dq(\phi)}{d\phi} \quad (1)$$

Flux controlled memristor.

$$v(t) = M(q(t))i(t), M(q) \equiv \frac{d\phi(q)}{dq} \tag{2}$$

Charge controlled memristor.

The differential equations describe the general behavior of a memristor, but a lot of details are missing as a real physical memristor is more complex. Real memristor devices often exhibit current-gated or voltage-gated behavior, in which voltage and current below a certain threshold do not change the memristance of the device. To simulate this behavior, more complex mathematical models are used. Notably, the ThrEshold Adaptive Memristor (TEAM) model [6] and the Voltage TEAM (VTEAM) model [7] are widely used in SPICE and custom circuit simulations. These models provide insights into the underlying physics mechanisms and possibly extrapolation of device behavior under unknown circumstances.

Apart from analytical models, phenomenological models are hardware-calibrated and meant to capture the nuances of memristor behavior in practical applications. The phase-change memory (PCM) model by Nandakumar et al. [8] calibrated on 10,000 PCM devices, and the metal-oxide memristor model by Nili et al. [9] is calibrated on 324 devices at six different temperatures. These models provide accurate and polynomial time interpolation of device non-idealities, which are ideal for large-scale integration and security applications. In **Figure 2**, we show the voltage-current relationship of 200 metal-oxide memristor devices initialized at 65 S, simulated with the Python implementation of Nili's model.

3.3 Simulation of memristor system

Memristors are not used in isolation but are often integrated into more extensive systems and circuits. These systems may include neuromorphic computing architectures, non-volatile memory arrays, or analog computing devices. To gain a comprehensive understanding of how memristors behave in practical applications, it is essential to simulate entire memristor-based systems.

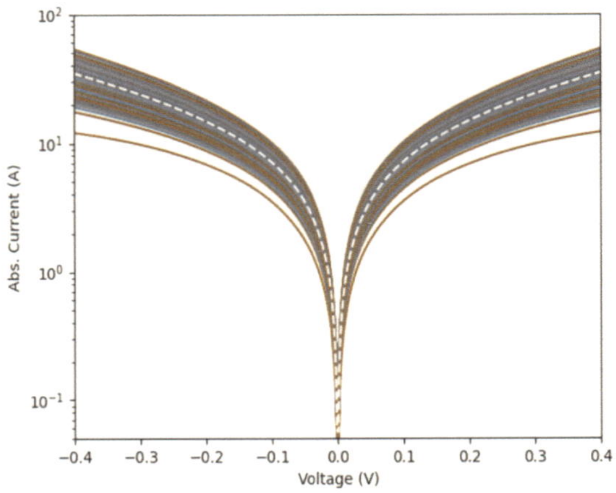

Figure 2.
Voltage-current relationship of 200 metal-oxide memristor devices initialized at 65S, operating at 100 MHz and 85°C. Variation is due to device-to-device variability, non-linearity, and operating noise.

Simulating memristor systems involves not only modeling individual memristor devices but also integrating these models with other circuit components. In general, memristors are arranged in a crossbar fashion to perform vector–matrix multiplications (VMM). Fully passive (0T1R) or 1-Transistor 1-Resistor (1T1R) schemes are often used architectures, with circuit complexity, energy efficiency, and non-idealities trade-offs. As shown in **Figure 3**, 0T1R arrangement has memristors being sandwiched between word lines and bit lines, whereas the 1T1R arrangement has source lines allowing the selection of specific memristors.

Compared to 1T1R arrangement, 0T1R arrangement is more pruned to line resistance, which refers to the resistance introduced by the crossbar word line and bit lines. The direct impact is voltage drop, as illustrated in **Figure 4**. To simulate this behavior, Chen proposed an iterative matrix equation solution [10] that is compatible with devices with non-linear IV relationships. The inclusion of this line resistance is important, as it may dominate the non-idealities at large-scale integration. To illustrate this, we implemented Nili's device model with Chen's crossbar line resistance model. In **Figure 4**, we show the impact of line resistance on a 128 × 128 0T1R memristor crossbar under (1) voltage applied from a single side, (2) voltage applied from both sides, and (3) tilization with 16 × 16 arrays.

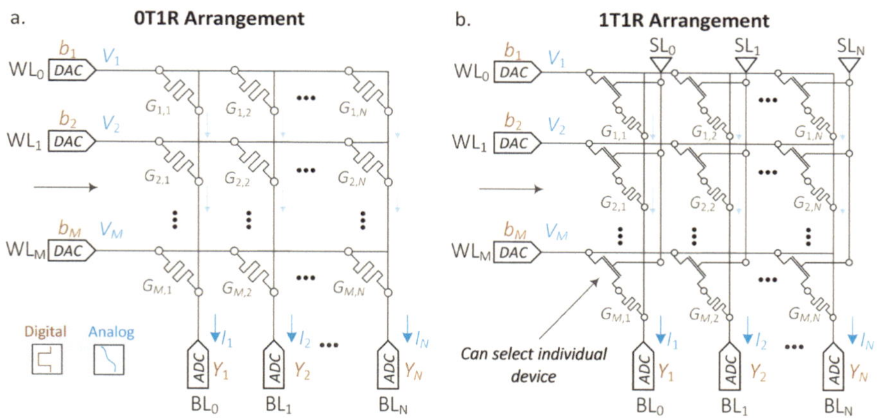

Figure 3.
Illustration of (a) 0T1R crossbar architecture and (b) 1T1R crossbar architecture. VMM can be performed with voltage encoding feed in from the word lines (WLs) and output read from the bit lines (BLs). In the 1T1R arrangement, individual devices can be selected using the select lines (SLs).

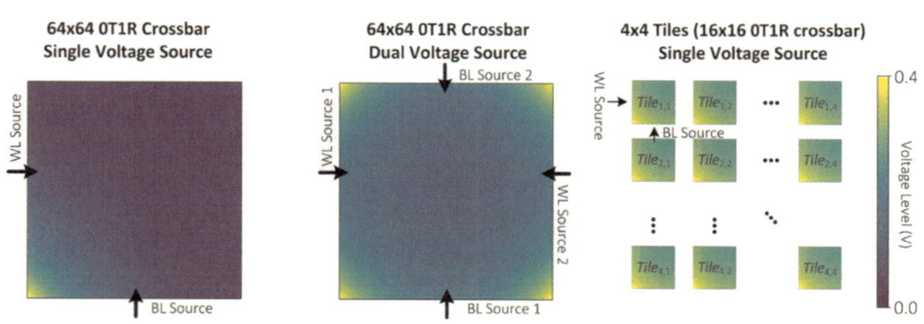

Figure 4.
Voltage drop due to line resistance on a 64 × 64 0T1R crossbar initialized with 65 μS mean conductance, with 20 Ω line resistance. (a) Cell voltage when input is delivered from a single side of the word lines. (b) Cell voltage when input is delivered from both sides of the word lines. (c) Cell voltage in tilized 16 × 16 arrays with single-sided word line voltage delivery.

3.4 Applications in deep learning

One of the most computationally demanding uses of memristor simulation is deep learning. Unlike other applications, deep learning models contain millions of parameters represented by the conductance states of the memristors. Therefore, simulation fidelity, alongside speed and memory consumption, are important metrics. Instead of creating a simulation framework from scratch, there are established open-source frameworks available.

The IBM Analog Hardware Acceleration Kit (IBM AIHWKit) [11] is an open-source Python/C++/CUDA toolkit designed to simulate in-memory computing devices with compatibility with the PyTorch ML framework. IBM AIHWKit supports on-chip training with in-memory gradient computation and inference with real device non-idealities based on statistical models and circuit non-idealities models.

MemTorch [12] is a similar toolkit to IBM AIHWKit, with a simpler VTEAM device model and 1T1R, 0T1R tilized memristor arrays. MemTorch supports numerous simulations of device and circuit non-idealities, including device-to-device variability, finite conductance states, and failure toward high/low resistance states. Contrary to IBM AIHWKit, which aims to provide an accurate simulation, MemTorch allows the exploration of noise levels beyond the usual operating range. In later sections, we will evaluate several proposed non-ideality mitigation methods using the MemTorch library to show robustness under hypothetical scenarios.

NeuroSim [13] is another framework for memristive deep neural network simulation built on C++. Compared to the aforementioned toolkits, NeuroSim provides simulation validated through real silicon devices and accurate power and footprint analysis. The framework is interfaced with the PyTorch library through DNN + NeuroSim [14], enabling an end-to-end benchmark framework for memristive deep neural network simulation. We provide a comparison of the three frameworks in **Table 1**.

4. Methods to mitigate memristor non-idealities

There are two classes of methods that aim to mitigate the effects of non-idealities in memristors. The first is the classical approach, using hardware-based solutions that aim to tackle a specific type of non-ideality for a specific type of device. The second is

Framework	IBM AIHWKit [11] 2021	MemTorch [12] 2020	DNN + NeuroSim [14] 2020
Prog. lang.	Python, C++, CUDA	Python, C++, CUDA	Python, C++
ML framework	PyTorch	PyTorch	PyTorch, Tensorflow
Device model	Statistical	Analytical	Statistical
In-situ training	✓	✓	✓
In-situ grad. comp.	✓		
Perf. estimation			✓

Table 1.
Comparison of open-source memristive deep neural network simulation frameworks.

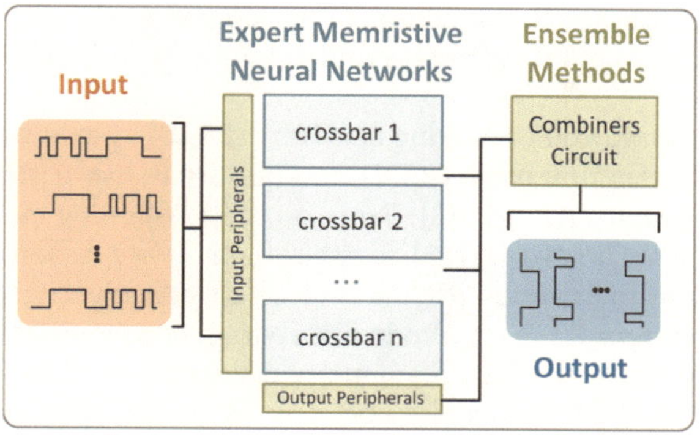

Figure 5.
Memristive deep neural network system with ensemble methods.

a relatively new but more promising approach, which aims to find general solutions that are applicable for all kinds of non-idealities; these are applicable in the context of memristive deep neural networks (MDNNs-DNNs implemented on memristor crossbars) and make use of techniques from machine learning such as ensemble methods, which combine the outputs of multiple individual networks to produce a more accurate overall output [15]; see **Figure 5** for a depiction of this second class of methods. In this section, we will first go over the first class of methods by analyzing some of these methods and discussing the pros and cons, and then we will go over the second class of methods.

4.1 Some hardware-based attempts to deal with the effects of non-idealities in memristors

The first solution we will be looking at is from the 2017 paper by Xia et al., which targets the device faults non-ideality in Resistive Random Access Memory (RRAM)-based Computing Systems (RCSs) by using redundancy schemes and a modified mapping algorithm.

This paper presents a fault-tolerant framework for RCSs [16]. It introduces a mapping algorithm with inner fault tolerance (MAO) designed to convert matrix parameters into RRAM conductances, effectively managing Stuck-At-Faults (SAFs) by optimizing the mapping process. Additionally, this paper proposes two baseline redundancy schemes to enhance RCS's effectiveness when a significant percentage of RRAM cells are faulty. To further adapt to varying SAF distributions and reduce the number of redundant RRAM cells, this paper introduces a distribution-aware redundancy scheme and a re-configurable redundancy scheme. These schemes provide dynamic fault tolerance capabilities. In essence, the paper proposes a comprehensive approach to address SAFs in RRAM Computing Systems, encompassing fault-tolerant mapping algorithms and adaptive redundancy schemes, ultimately improving the reliability and accuracy of RRAM-based computations.

The second solution we will be looking at is from the 2018 paper by Ambro-gio et al., which targets the limited dynamic range non-ideality in PCM-based memristive devices by using sets of varying conductances for each synaptic weight [17]. This paper presents mixed hardware-software neural network implementations, incorporating PCM (phase-change memory) for long-term storage, volatile capacitors for near-linear updates, and weight-data transfer with "polarity inversion" to mitigate device variations.

The third solution we will be looking at is from the 2018 paper by Chai et al., which targets the random telegraph noise (RTN) non-ideality in memristive devices by adopting non-filamentary RRAM [18]. Based on their switching mechanisms, memristive devices can be categorized as either filamentary or non-filamentary. The authors find that non-filamentary RRAM devices, specifically TiO2/a-Si (a-VMCO), exhibit a more consistent RTN amplitude distribution and significantly lower RTN occurrence rates compared to their filamentary counterparts, thus suggesting that non-filamentary RRAM devices can be a good choice for mitigating the effects of the RTN non-ideality in memrstive systems.

The fourth solution we will be looking at is from the 2018 paper by Fang et al., which targets the device-to-device variability non-ideality in HFOx RRAM memristive devices by introducing an ultra-thin ALD-TiN buffer layer [19]. This paper investigates the impact of oxygen-scavenging metal, specifically titanium (Ti), on the performance of HfOx-based Resistive Random Access Memory (RRAM) devices. Through experimentation, the researchers found that during the physical vapor deposition process, Ti atoms could penetrate randomly into the bulk of HfOx, leading to the creation of oxygen vacancies and subsequent performance degradation in RRAM devices. To address this issue, the paper proposes a novel structure involving the insertion of an atomic layer deposition (ALD) TiN buffer layer between the oxygen-scavenging Ti layer and the HfOx layer. This buffer layer effectively reduces the negative effects of Ti penetration and enhances the vacancy-creating process. Experimental results confirmed that incorporating the TiN buffer layer significantly improved the uniformity of switching parameters, such as forming voltage and resistance, in RRAM devices.

4.2 Ensemble methods for mitigating non-idealities

4.2.1 Ensemble Averaging

The most basic type of ensemble method is known as Ensemble Averaging (EA). In this method, the output of multiple neural networks is simply averaged (linearly, meaning that each network is weighed the same regardless of individual performance) to produce a combined output. Each of the individual networks comprising the ensemble is trained on the complete dataset. This method can be summarized with the following equation:

$$O(\mathbf{x}) = \frac{1}{N}\sum_{i=1}^{N} M_i(\mathbf{x}) \tag{3}$$

Here, $\boldsymbol{x}$ is the input, $M_i(\mathbf{x})$ is the output of network i for input $\boldsymbol{x}$, N is the total number of networks in the ensemble, and $O(\boldsymbol{x})$ is the output of the ensemble for input $\boldsymbol{x}$.

4.2.2 Generalized Ensemble Method

The Generalized Ensemble Method (GEM) is a method that is similar to EA, with the only main difference being that the average is not linear and instead, networks are weighed differently; there is not any one specific scheme that is used to determine how each network is weighed and the most optimal scheme can differ depending on the application and other details. In our work in (cite ensemble method paper), we found a weighting scheme that worked best for the basic task of image classification on the MNIST dataset; please see that paper for more details regarding this. The equation summarizing this method is as follows, with the notation defined similarly as before, the only new addition being that w_i is the weight of network i, determined according to whatever weighting scheme is used:

$$O(\mathbf{x}) = \frac{\sum_{i=1}^{N} w_i M_i(\mathbf{x})}{\sum_{i=1}^{N} w_i} \tag{4}$$

4.2.3 Voting

Unlike the previous two ensemble methods, voting is only applicable to classification tasks. The most common type of voting, known as plurality voting, is as follows. Each individual network in the ensemble independently makes a prediction from the input, and the class that is most commonly predicted is chosen as the output of the ensemble. Ties can be broken arbitrarily or using a heuristic (source for this?). This method can be summarized by the following equation:

$$O(\mathbf{x}) = c_{argmax_j \sum_{i=1}^{N} o_i^j(\mathbf{x})} \tag{5}$$

Here c_k represents the classes and $o_i^j(\mathbf{x})$ represents the output of network i for class c_j given input $\boldsymbol{x}$, which takes the value 1 if the prediction of the network is c_j and 0 otherwise.

4.2.4 Weighted Voting

In the same way that GEM is an extension of EA, Weighted Voting is an extension of voting in the sense that instead of the vote of each network in the ensemble being weighed equally regardless of their individual performance, the networks can be weighed differently, meaning that, for example, the vote of a better performing network could take more precedence over that of a poorer performing network. There are many possible weighting schemes that can be used for this method. The following is a derivation of what is the theoretically optimal weighting scheme (given some assumptions) [20]. Let the outputs of the individual networks in the ensemble for an input $\boldsymbol{x}$ be $\mathbf{1} = (l_1, l_2, \ldots, l_N)^T$ where N is the total number of networks in the ensemble, and l_i is the label of the class predicted by network i. The following is the Bayesian optimal discriminant function for the combined output of the ensemble on the class with label c_j:

$$H^j(\mathbf{x}) = log\left(P(c_j)P(\mathbf{1}|c_j)\right) \tag{6}$$

Now, if we make the assumption that the outputs of each network are conditionally independent, meaning that $P(\mathbf{1}|c_j) = \prod_{i=1}^{N} P(l_i|c_j)$, then the Bayesian optimal discriminant function can be simplified as follows:

$$\begin{aligned} H^j(\mathbf{x}) &= \log(P(c_j)) + \log\left(\prod_{i=1}^{N} P(l_i|c_j)\right) \\ &= \log(P(c_j)) + \log\left(\prod_{i=1,\, l_i=c_j}^{N} P(l_i|c_j) \prod_{i=1,\, l_i\neq c_j}^{N} P(l_i|c_j)\right) \\ &= \log(P(c_j)) + \log\left(\prod_{i=1,\, l_i=c_j}^{N} p_i \prod_{i=1,\, l_i\neq c_j}^{N} (1-p_i)\right) \\ &= \log(P(c_j)) + \sum_{i=1,\, l_i=c_j}^{N} \log\left(\frac{p_i}{1-p_i}\right) + \sum_{i=1}^{N} \log(1-p_i) \end{aligned} \tag{7}$$

Here, p_i denotes the accuracy of network i. Note that the final term in our simplified expression does not depend on c_j, and thus can be ignored. Moreover, the first term does not depend on the individual networks, so from the second term, we see that the theoretically optimal weighting scheme for Weighted Voting would satisfy the following:

$$w_i \propto \log\left(\frac{p_i}{1-p_i}\right) \tag{8}$$

In practice, it turns out that due to the assumptions we made to derive this result that is not satisfied in reality (the assumptions being that the outputs of each network are conditionally independent and that we know the ground truth accuracies of the networks), there are other weighting schemes that can result in a greater improvement in performance depending on the task at hand; the reader is referred to our work in [21] for some more details regarding this for the specific task of handwritten digit recognition in the MNIST dataset.

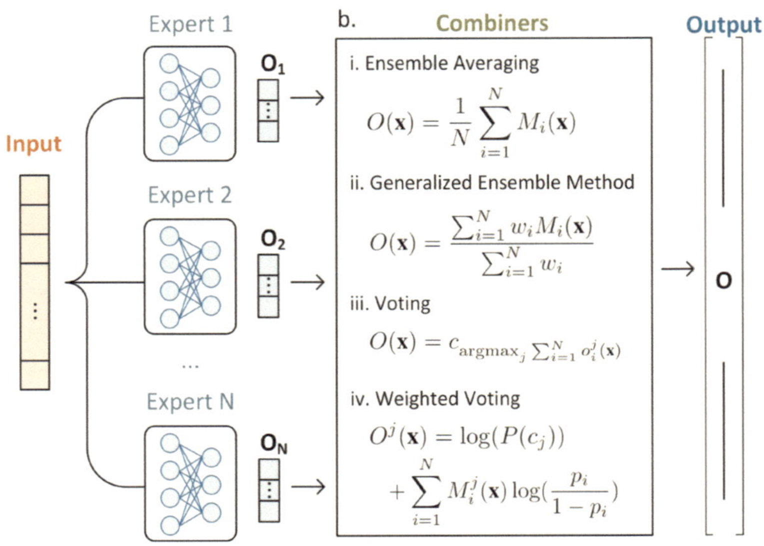

Figure 6.
A summary of ensemble methods techniques: four different ways to combine outputs of multiple experts introduced in Section 4.2.

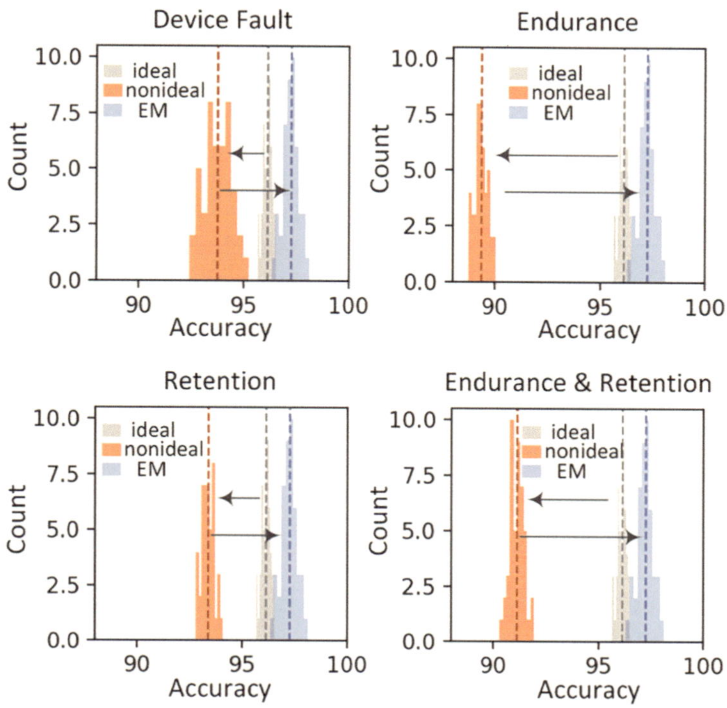

Figure 7.
Performance of memristive deep neural networks (MDNN) for the MNIST hand digit recognition dataset under various non-idealities: device fault, endurance, retention, and endurance & retention combined. Distribution gathered from 10 trials. Ensemble methods (EM) combined outputs from five experts.

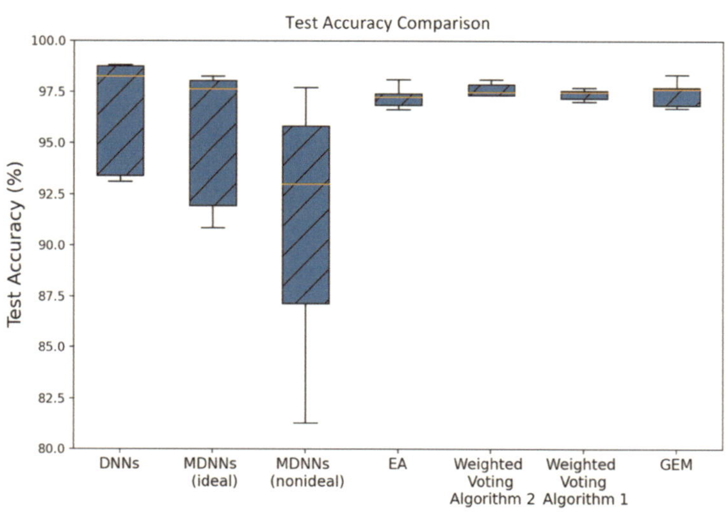

Figure 8.
Performance comparison of four ensemble methods techniques and baseline ideal/non-ideal memristive deep neural networks (MDNNs) for the MNIST hand digit recognition dataset. Distribution gathered from 10 trials. Ensemble methods (EM) combined outputs from five experts.

4.2.5 Ensemble methods: summary

In summary, we introduced four ensemble methods that mitigate MDNN non-idealities. We summarized the four techniques in **Figure 6**. We evaluated the performance of ensemble methods on the handwritten digit recognition task with the MNIST dataset, with experiments gathered from 10 trials, and each ensemble method combined outputs from five different experts [21]. As a brief summary of our work, **Figure 7** demonstrates the performance of the ensemble method compared to ideal and non-ideal MDNNs under device fault, endurance, retention, and endurance and retention combined non-idealities. **Figure 8** compares MDNN accuracy against the baseline and across different ensemble methods. The reader is again referred to our work in [21] for detailed implementation of these ensemble methods.

5. Conclusions

In conclusion, this chapter has explored various classes of memristor non-idealities, ways to simulate them, and hardware and universal ways to address them. By conducting simulations, we assessed the effects of diverse non-idealities and proposed strategies for their mitigation with a focus on memristive deep neural networks. This chapter furnishes readers with practical guidance for mitigating non-idealities, facilitating the design of more robust, more efficient memristor circuits and systems.

Author details

Muhammad Ahsan Kaleem[1†], Jack Cai[1†], Yao-Feng Chang[2], Roman Genov[1] and Amirali Amirsoleimani[3*]

1 University of Toronto, Toronto, Canada

2 Intel Corporation, United States of America

3 York University, Toronto, Canada

*Address all correspondence to: amirsol@yorku.ca

† These authors contributed equally.

References

[1] Li S et al. Wafer-scale 2D hafnium diselenide based memristor crossbar array for energy-efficient neural network hardware. Advanced Materials. 2021;**34**(25):2103376. DOI: 10.1002/adma.202103376 [Accessed: November 5, 2023]

[2] Chen S et al. Wafer-scale integration of two-dimensional materials in high-density memristive crossbar arrays for artificial neural networks. Nature Electronics. 2020;**3**(10):638-645. DOI: 10.1038/s41928-020-00473-w [Accessed: November 5, 2023]

[3] Kumar S, Gautam MK, Yadav S, Mukherjee S. Memcapacitive to memristive transition in Al/Y2O3/GZO crossbar array. IEEE Transactions on Electron Devices. 2023;**70**(6):3341-3346. DOI: 10.1109/TED.2023.3265622 [Accessed: November 5, 2023]

[4] Kumar S, Agarwal A, Mukherjee S. Electrical performance of large-area Y_2O_3 memristive crossbar array with ultralow C2C variability. IEEE Transactions on Electron Devices. 2022;**69**(7):3660-3666. DOI: 10.1109/TED.2022.3172400 [Accessed: November 5, 2023]

[5] Chua L. Memristor-the missing circuit element. IEEE Transactions on Circuit Theory. 1971;**18**(5):507-519. DOI: 10.1109/TCT.1971.1083337 [Accessed: September 29, 2023]

[6] Kvatinsky S, Friedman EG, Kolodny A, Weiser UC. TEAM: ThrEshold adaptive memristor model. IEEE Transactions on Circuits and Systems I: Regular Papers. 2013;**60**(1):211-221. DOI: 10.1109/TCSI.2012.2215714 [Accessed: September 29, 2023]

[7] Kvatinsky S, Ramadan M, Friedman EG, Kolodny A. VTEAM: A general model for voltage-controlled memristors. IEEE Transactions on Circuits and Systems II: Express Briefs. 2015;**62**(8):786-790. DOI: 10.1109/TCSII.2015.2433536 [Accessed: September 29, 2023]

[8] Nandakumar SR et al. Phase-change memory models for deep learning training and inference. In: 2019 26th IEEE International Conference on Electronics, Circuits and Systems (ICECS), Genoa, Italy. Genoa, Italy: IEEE; 2019. pp. 727-730. DOI: 10.1109/ICECS46596.2019.8964852 [Accessed: September 29, 2023]

[9] Nili H, Vincent AF, Prezesio M, Mahmoodi MR, Kataeva I, Strukov DB. Comprehensive compact phenomenological modeling of integrated metal-oxide memristors. IEEE Transactions on Nanotechnology. 2020;**19**:344-349. DOI: 10.1109/TNANO.2020.2982128 [Accessed: September 29, 2023]

[10] Chen A. A comprehensive crossbar Array model with solutions for line resistance and nonlinear device characteristics. IEEE Transactions on Electron Devices. 2013;**60**(4):1318-1326. DOI: 10.1109/TED.2013.2246791 [Accessed: September 29, 2023]

[11] Rasch MJ et al. A flexible and fast PyTorch toolkit for simulating training and inference on analog crossbar arrays. In: 2021 IEEE 3rd International Conference on Artificial Intelligence Circuits and Systems (AICAS). Washington DC, DC, USA: IEEE; 2021. pp. 1-4. DOI: 10.1109/AICAS51828.2021.9458494 [Accessed: September 29, 2023]

[12] Lammie C, Azghadi MR. MemTorch: A simulation framework for deep

Memristive cross-Bar architectures. In: 2020 IEEE International Symposium on Circuits and Systems (ISCAS). Vol. 2020. Seville, Spain: IEEE; 2020. pp. 1-5. DOI: 10.1109/ISCAS45731. 2020. 9180810 [Accessed: September 29, 2023]

[13] Lu A, Peng X, Li W, Jiang H, Yu S. NeuroSim simulator for compute-in-memory hardware accelerator: Validation and benchmark. Frontiers in Artificial Intelligence. 2021;**4**:659060. DOI: 10.3389/frai.2021.659060 [Accessed: September 29, 2023]

[14] Peng X, Huang S, Jiang H, Lu A, Yu S. DNN+NeuroSim V2.0: An End-to-End Benchmarking Framework for Compute-in-Memory Accelerators for On-chip Training. 2020. Available from: https://arxiv.org/abs/2003.06471 [Accessed: September 29, 2023]

[15] Joksas D, Freitas P, Chai Z, et al. Committee Machines—A Universal Method to Deal with Non-idealities in Memristor-based Neural Networks. 2020. Available from: https://www.nature.com/articles/s41467-020-18098-0 [Accessed: September 30, 2023]

[16] Xia L, Huangfu W, Tang T, Yin X. Stuck-at Fault Tolerance in RRAM Computing Systems. 2017. Available from: https://ieeexplore.ieee.org/ielaam/5503868/8330765/8119491-aam.pdf [Accessed: September 30, 2023]

[17] Ambrogio S, Narayanan P, Tsai H, et al. Equivalent-accuracy Accelerated Neural-Network Training Using Analogue Memory. 2018. Available from: https://europepmc.org/article/med/29875487 [Accessed: September 30, 2023]

[18] Chai Z, Freitas P, Zhang W, Hatem F, Fu Zhang J, Marsland J, et al. Impact of RTN on Pattern Recognition Accuracy of RRAM-Based Synaptic Neural Network. 2018. Available from: https://www.researchgate.net/publication/327484337 [Accessed: September 30, 2023]

[19] Fang Y et al. Improvement of HfOx-Based RRAM Device Variation by Inserting ALD TiN Buffer Layer. 2018. Available from: https://ieeexplore.ieee.org/document/8352781 [Accessed: September 30, 2023]

[20] Zhou Z-H. Ensemble Methods: Foundations and Algorithms. 2012. [Accessed: September 30, 2023]

[21] Kaleem MA, Cai J, Amirsoleimani A, Genov R. A Survey of Ensemble Methods for Mitigating Memristive Neural Network Non-idealities. 2023. Available from: https://ieeexplore.ieee.org/document/10181553 [Accessed: September 30, 2023]

Chapter 9

Neuromorphic Computing with Resistive Memory and Bayesian Machines

Nikolay Frick

Abstract

Bio-inspired computing with memristors and neuristors offers promising pathways to energy-efficient intelligence. This work reviews toolkits for implementing spiking neural networks and Bayesian machine learning directly in hardware using these emerging devices. We first demonstrate that normally passive memristors can exhibit neuristor-like oscillatory behavior when heating and cooling is taken into account. Such oscillations enable spike-based neural computing. We then summarize recent works on leveraging intrinsic switching stochasticity in memristive devices to physically embed Bayesian models and perform in-situ probabilistic inference. While still facing challenges in endurance, variation tolerance, and peripheral circuitry, this co-design approach combining tailored algorithms and nanodevices could enable a new class of ultra-low power brain-inspired intelligence tolerant to uncertainty and capable to learn with small datasets. Longer-term, hybrid CMOS-memristor systems with sensing/actuation may provide fully adaptive Bayesian edge intelligence. Overall, the confluence of probabilistic algorithms and memristive hardware holds promise for future electronics combining efficiency, adaptability, and human-like reasoning. Academic innovations exploring this algorithm-hardware co-design can lay the foundation for this emerging paradigm of probabilistic cognitive computing.

Keywords: memristor, neuristor, Bayesian machine, stochastic, neuromorphic

1. Introduction

Brain is one of the most complex objects known to humanity with most of it's functionality yet to be discovered. The complexity of brain lays in it's self-organization that goes on through lifetime where individual neurons constantly seek new connections with proximate and distant neurons growing new and destroying old and unused synaptic connections. Each neural cell is surrounded by an environment full of ions with neuron's membrane serving as a barrier preventing an uncontrolled charge to flow through it. Instead the ion channels on the surface of the neuron opens and closes at specific times when concentration of particular charge exceeds membrane threshold that results in membrane depolarization, causing transmission of spiking signal over synapse as shown in **Figure 1**. The spiking signal is then converted

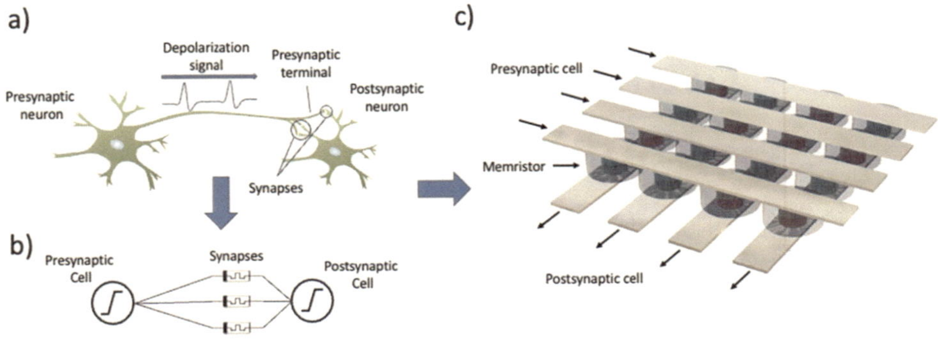

Figure 1.
Biological network and it's approximation with memristors. (a) Biological network with presynaptic and postsynaptic neurons with a spiking cell depolarization signal traveling along axon (image from https://scidraw.io/drawing/610). (b) An approximated artificial network with memristors acting as synapses. (c) Crossbar array with artificial memristive synapses as an efficient approach to neuromorphic computation.

into chemical signal via combination of neuromediators and transmitted further to postsynaptic cell. The frequency and density of the stimulation from presynaptic neuron affects the probability of activation of postsynaptic neuron, whereas this neuron contributes to signal propagation to other downstream neurons.

In biological neurons excitation from one neuron propagates to downstream neurons in a branching fashion, frequently with probabilities below 1, which is a signature of subcritical interaction [1]. In other words it is unlikely that signal from a single neuron will be noticed by large number of neighbor neurons and will frequently result in a random-walk-like pathway that fades away after few dozens of cells but can propagate for many more longer steps. This subcritical interaction between neurons, facilitates the highest possible efficiency of information transmission in biological neural network [2] and increases it's dynamic range [3].

Configuring networks of nonlinear neuron-like elements enables efficient and flexible parallel analog computation, which helps biological organisms survive by solving complex problems such as finding food and avoiding threats [4]. In particular, the nonlinear interactions between neurons are facilitated by the brain's adaptive plasticity, allowing synapses to strengthen or weaken [5]. This synaptic plasticity enables a variety of arithmetical operations on neural signals [6], including multiplication, addition, subtraction, convolution, and more [7]. By leveraging similar principles, artificial neural networks aim to achieve brain-like flexibility and computational power.

Mathematically, a biological neural network can be approximated as a nonlinear function that accepts any number of weighted inputs from upstream neurons [4]. This function applies a threshold to sum the inputs, converting them into a scalar neuronal activation value (**Figure 1**) [8]. Even simple artificial neural network models based on this principle, using synapses as real-valued connection weights and neurons as summators with thresholding transfer functions, have achieved tremendous successes in applications like image recognition [9], natural language processing [10], and content generation [11]. The ability of artificial neural networks to perform complex information processing tasks has sparked a revolution in the information era [12]. By emulating aspects of brain computation, techniques like deep learning are powering advances in fields from computer vision to healthcare [13].

However, modern artificial neural networks only imitate a fraction of biological neural network functionality, primarily neuron activation functions and synaptic weight connections. An interesting technique borrowed from biology is randomly "dropping out" weights during training, which regularizes artificial networks by improving their robustness [14, 15]. This weight dropout technique is surprisingly effective, and is now widely used when training neural networks across many applications, including large language models like GPT [16]. Randomly disabling connections seems to mimic the noisy yet fault-tolerant nature of biological neural processing [14]. While artificial networks do not yet match the full complexity of biological computation, techniques like weight dropout demonstrate how we can take inspiration from neuroscience to improve deep learning.

While deep artificial neural networks have achieved remarkable results on narrow tasks, their simple neuron models and matrix multiply operations remain primitive compared to biological brains. Modern deep learning approaches lack the adaptability, fault tolerance, and uncertainty handling of actual neural circuits. Their massive data requirements and susceptibility to unintended bias also contrast the efficiency and versatility of biological learning. Additionally, approximations of spiking dynamics in artificial networks are quite basic compared to rich temporal processing in real neurons. Unlike fault-tolerant biological hardware, artificial neural network implementations are not inherently robust. These limitations motivate exploring alternative bio-inspired computing paradigms. This work reviews recent progress on leveraging memristive devices to realize more efficient and robust neural computing. We explore the active properties arising in passive memristor networks, which enable spiking oscillations similar to neuristors. We also discuss implementing Bayesian inference directly in memristive hardware, achieving efficient learning from limited data while avoiding overfitting. By emulating additional facets of biological neural computation, from spiking dynamics to probabilistic reasoning, memristor-based architectures could provide a path towards more adaptive, resilient, and low-power brain-inspired intelligence.

2 Neuromorphic computing with memristors and neuristors

2.1 Memristors

Memristors are passive circuit elements that exhibit resistance based on the history of applied voltage or current. This memristance arises in devices with mobile cationic or anionic type of carriers that drift within an insulating layer under an electric field, altering the internal resistance [17, 18]. For example, in cation-based memristors, such as chalcogenides Ag_2S or Cu_2S applying a positive voltage causes Ag^+ or Cu_{2^+} ions to drift towards the cathode, forming conductive bridges that decrease resistance [19]. Similarly, reversing the polarity of the voltage causes break-down the conductive filaments redusing the conductivity of the device. The memristance depends on the extent of ionic transport within the device [20]. Compared to CMOS transistors, memristors offer improved energy efficiency and density, with switching energies down to 1 fJ per bit, and can retain non-volatile memory states even after power is removed [21]. However, switching speeds are often slower than solid-state electronics, ranging from nanoseconds up to milliseconds for full ion migration [22]. Further advances in materials and integration will help unleash the potential of memristors for low-power neuromorphic computing.

One widely studied memristor material is titanium dioxide (TiO_2) thin films, which exhibit voltage-controlled transport of oxygen vacancies within the oxide layer to alter resistance states [23, 24]. For example, in Pt/TiO_2/Pt devices, applying a positive voltage causes O_{2-} ions to drift towards the anode, creating oxygen vacancy rich filaments that locally increase conductivity [25]. Reversing the bias causes annihilation of these Magnéli filaments, increasing the device resistance [26]. The time for switching between the high and low resistance states in TiO_2 memristors is typically on the order of 100 ns for filament formation or rupture [27]. However, the full timescale over which the memristance evolves can range from microseconds for small perturbations up to seconds for reaching extremal resistances, depending on voltage amplitude, film thickness, and device structure [28, 29]. Further research on interfaces and doping seeks to improve the stability and tunability of TiO_2-based memristors for neuromorphic computing applications.

The memristance of a basic memristive device can be modeled as a thin film with thickness D, as described by Strukov et al. [18]. The memristance M depends on the internal state variable x, which represents the position of dopants or effective filament length within the device shown in Eqs. (1) and (2).

Here, R_{ON} is the low resistance state when x = 1 corresponding to maximum doping or full filament formation. R_{OFF} is the high resistance state when x = 0 corresponding to no doping or an open filament gap. The internal state variable x evolves according to Eq. (2) where F is a function describing the rate of filament growth or dopant drift under an applied voltage V and time t resulting in current I. Eqs. (1) and (2) link the internal material dynamics to the externally measurable memristance M, which can vary continuously between R_{ON} and R_{OFF} depending on the value of x. To simplify numerical integration, a window function is added, that is denoted by F(x). This basic model captures the essence of resistance switching in memristive devices based on filamentary or dopant migration mechanisms.

$$M = R_{on}\frac{x(t)}{D} + R_{off}\left(1 - \frac{x(t)}{D}\right) \tag{1}$$

$$\frac{dx}{dt} = \mu\frac{R_{on}}{D}IF(x) \tag{2}$$

While an individual memristor acts as a passive device, interconnecting memristors in networks can lead to remarkable emergent properties. As shown in **Figure 2**, we simulated a network of randomly oriented memristive nanowires governed by Eqs. (1) and (2). When subjected to oscillating voltage patterns, this network exhibited negative differential resistance (NDR)-meaning the current actually decreased in some voltage ranges. This NDR normally only occurs in active devices undergoing metal–insulator transitions. It enables oscillatory signal generation and neuristor-like spiking behavior. The emergence of NDR in these simulations suggests that networks of passive memristors can display unusual active characteristics. This arises from the complex interactions within the network topology. While individual memristors obey simple dynamics, connecting them together leads to more complex phenomena like NDR.

The emergent active properties arising in networks of passive memristors are intriguing, there remain open questions about fully characterizing and exploiting such effects for neuromorphic computing. However, progress has also been made in engineering truly active memristive devices that functionally mimic spiking oscillations in

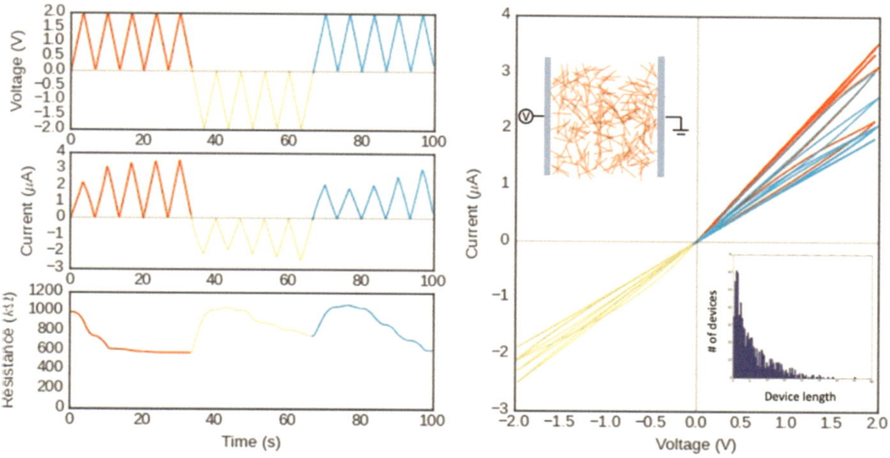

Figure 2.
Triangular voltage sweeps through a network of 879 randomly distributed memristors exhibiting a pattern of negative differential resistance.

biological neurons. These "neuristor" devices incorporate additional structures and physical effects beyond basic memristance to achieve neuron-like dynamics, moving beyond simply interconnecting passive memristors.

2.2 Neuristors

Neuristor is another class of two-terminal electronic devices that aim to functionally mimic dynamics of biological neuron. Neuristor was conceptualized in 1960s by Crane et al., that proposed a computation system with electrical elements exhibiting fatigue in order to imitate decay of charge propagation in neural soma [30, 31]. However, first attempts of electric implementation of neuristors was not scalable due to the need for inductors or bipolar thyristors [32, 33]. At around the same time, a novel threshold switching property of common oxides such as VO_2, VO and TiO was reported by Morin et al. [34] that showed an abrupt change in conductivity in these materials upon reaching particular temperatures as a result of metal to insulator transitions. More than half a century later this property was used by Pickett et al. to show the first neuristor device based on NbO_2, that demonstrated insulator-to-metal transitions that exhibited neuron-like oscillations without silicon transistors [35, 36] using only resistors and capacitors for internal state storage.

Another variant of neuristors was also reported based on VO_2 [37]. VO_2 is a termochromic material that undergoes an abrupt resistance switch around 340 K with characteristic color change due to a structural phase from semiconducting monoclinic to metallic rutile phase monoclinic [38, 39]. A typical I-V characteristics of a nanoscale devices shown in **Figure 3**. In this configuration, a small amount of nanoscale VO_2 material was sandwiched between two copper conductors in a 100 um diameter cavity. While triangular voltage sweep exhibited metal–insulator transitions between 10 and 20 V, the triangular current driven regime (y axis) showed negative differential resistance with characteristic.

Thin film VO_2 neuristors have exhibited spiking oscillations and threshold firing behavior analogous to action potentials in neurons [40–42]. The resistance transition

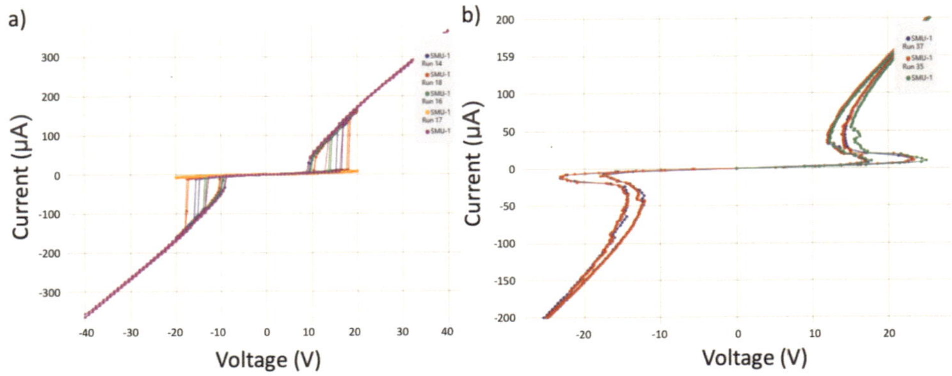

Figure 3.
Resistive switching of VO_2 device. (a) Triangular voltage sweep setup, showing metal–insulator phase transition (b) triangular current sweep setup, showing a characteristic negative differential resistance.

in these threshold switching materials can be induced thermally, providing a means to modulate neuronal spiking patterns without relying solely on electrical bias like traditional memristors.

Recently a caloritronic neuristor was reported which relies on the low temperature of metal–insulator phase transitions in VO_2 that allowed to eliminate capacitors needed to store internal state using only joule heating from a proximate resistor to move neuristor transition state closer to the phase transition or far away from it [43]. The reported device exhibited spiking with remarkable similarity to biological cells, namely Hodgkin-Huxley neuron model that opens opportunities for novel high-density neuromorphic computing [44].

Metal–insulator transitions enable spiking oscillations in neuristors, but they are not the only pathway. As we demonstrate, modifying the dynamics of plain memristors to include Joule heating and Newtonian cooling effects can also elicit oscillations. With carefully balanced thermal feedback, even simple memristive devices can exhibit self-sustaining spikes without undergoing a phase change.

Below we provide a model that could be used to simulate oscillations with a variant of NDR in a plain memristive nanowire with Joule heating and Newtonian cooling [45–47]. The parameters are depicted in **Table 1** while modified memristor equations with cooling and heating terms are provided in Eqs. (3) and (4).

$$R(w) = (R_{OFF} - (R_{OFF} - R_{ON}) * w/D) \tag{3}$$

$$\begin{aligned} \frac{dw}{dt} &= \mu \frac{R_{ON}}{D^2} i - \frac{1}{\tau}\frac{w}{D} - \frac{1}{10^{-4}}(T - T_c)\frac{w}{D} \\ \frac{dT}{dt} &= \frac{i^2}{\rho c \sigma A^2} - \frac{1}{10^{-4}}(T - T_c) \end{aligned} \tag{4}$$

The resulting state variable spiking signal generation is depicted in **Figure 4** where a rounded-corner square voltage pulse set across the device forces it to generate a packet of four spikes in the state variable w. Interestingly, that spiking is generated due to the interplay of cooling, heating and state change and is governed only by the system of two first order ordinary differential equations.

Even more interesting behavior of this system can be observed when driven by a periodic signal. In particular when driven by $\sin\left(\frac{2\pi t}{T}\right)$ the resulting oscillations become

R_{ON}	$1 \times 10^4\,\Omega$
R_{OFF}	$16 \times 10^5\,\Omega$
D	10 nm
μ	1×10^8 nm^2/s/V
ϵ	1×10^{-3}
w_0	0
T	0.01 s (period)
p	10
τ	1×10^2
ρ	10.5 kg/m^3
c	240 J/kg/K
σ	6.3×10^7 S/m
A	$2\pi(100 \times 10^{-9})^2$ m^2 (wire cross section area)
T_c	300 K

Table 1.
Memristor model parameters.

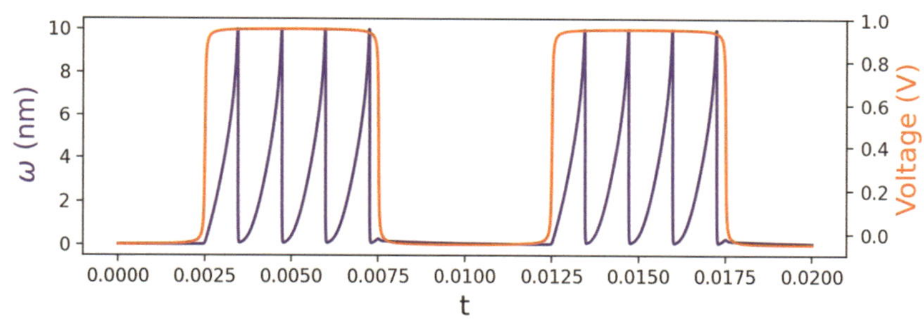

Figure 4.
Simulation of spiking in a memristive nanowire that undergoes simultaneous Newtonian cooling and Joule heating.

chaotic as shown in **Figure 5**. These temporal chaotic oscillations could be an alternative approach to stochastic computing as was proposed previously [48].

Further investigation is required to confirm these type of oscillations in passive memristive devices with some insights highlights to approaches previously reported elsewhere [19].

However, the underlying physics of neuristors and memristors both involve modulating material resistivity to achieve analog memory states. Combining these devices in crossbar arrays enables dense neuromorphic systems that leverage bio-inspired computing principles [49]. The development of neuristors expands the toolkit for hardware-based neural networks beyond standard memristive elements providing environment for neuromorphic computing.

Unlike artificial neural networks, which use simplified rate-based neurons and static synaptic weights, neuristors directly mimic the rich temporal spiking dynamics

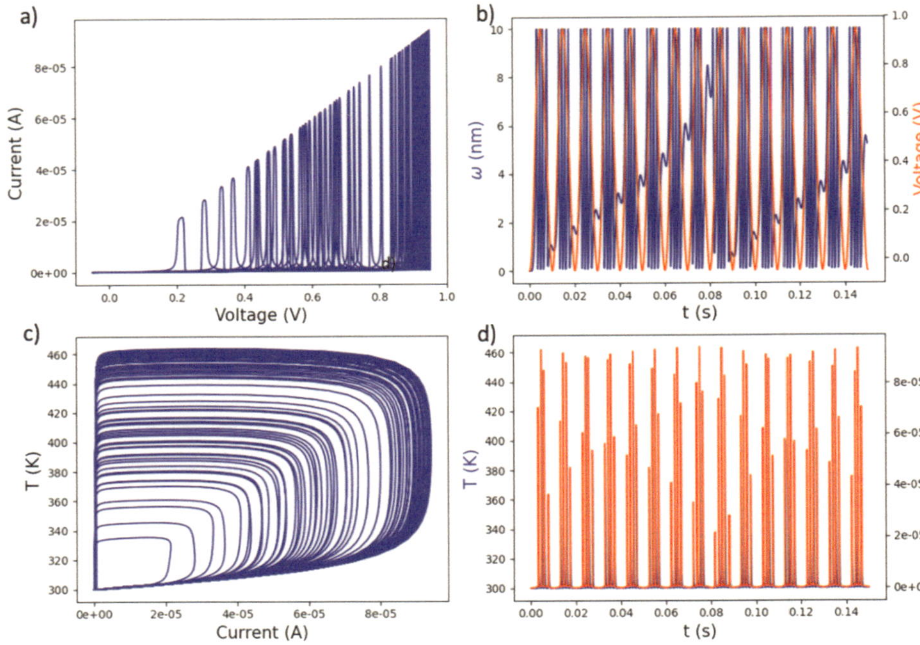

Figure 5.
Simulation of chaotic spiking in a memristive nanowire that undergoes simultaneous Newtonian cooling and Joule heating when driven with periodic voltage signal. (a) The IV characteristics of the device showing sporadic current spikes that result in rapid heating and followed by current drop due to the state switch (b) resulting change in state variable ω under influence of periodic voltage stimulation (c) the relationship between current and wire temperature (d) combined overview of current spiking and temperature as function of time.

of biological neurons. Rather than approximating spikes as averaged rates, neuristors exhibit self-sustaining oscillations and action potential-like firing patterns. This enables neuristor-based systems to capture the timing and rhythms of boinspired neuromorphic computation, which convey valuable information lost in rate-based Artificial Neural Network (ANN) models. As a result, neuristor networks have the potential to achieve brain-like capabilities beyond merely efficient statistical pattern recognition, such as complex temporal processing, real-time adaptability, and self-awareness [50]. While challenging to perfect, approximating both the architecture and temporal dynamics of biological cognition could enable revolutionary advances in artificial intelligence, beyond the constraints of feedforward deep learning. Looking forward, combining neuristor-based spiking dynamics with probabilistic modeling through Bayesian computing could provide the missing ingredients for fully brain-inspired intelligence.

2.3 Bayesian computing with stochastic memristors

While the goal for conventional neural networks is to learn a static weight, in Bayesian neural networks the weight is a distribution as shown in **Figure 6**. This is a one step towards biologically plausible neural networks as in biological synapses weights are modulated via chemical messengers across the synapse [7]. The complex balance and timing of different neurotransmitters binding to postsynaptic receptors thus encodes the weighting or strength of each synapse [51]. Therefore with their reliance on probabilistic inference and dynamic reweighting of connections, Bayesian

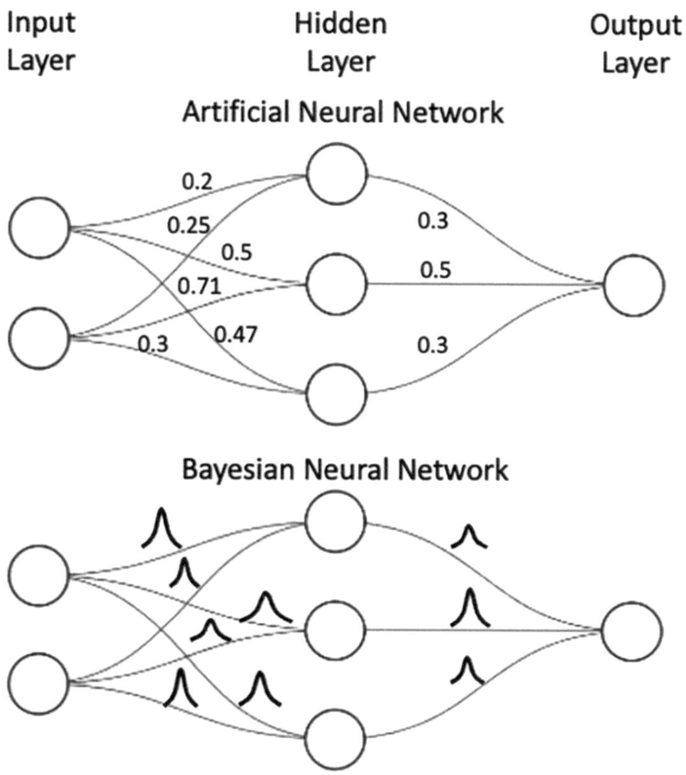

Figure 6.
Comparing ANNs to Bayesian Neural Networks (BNNs). ANNs store exact weights that makes them easy to overfit to data or being overconfident. BNNs on the other hand store weight distribution, that allows this network to learn from smaller datasets, while making them more robust and resistant to overfitting.

neural networks may more closely reflect the uncertainty, adaptability, and neurotransmitter-modulated plasticity inherent in biological neural circuits than conventional artificial neural networks with static synaptic weights.

Recently, Harabi and colleagues developed an innovative memristor-based architecture to perform Bayesian inference entirely in hardware [52]. The system stores likelihood values distributed across separate arrays of hafnium oxide memristors, leveraging their non-volatile memory capability. These likelihood arrays connect with digital logic to physically encode the topology of a Bayesian network. Each memristor is operated as a binary device, programmed to a high or low conductance state to represent a probability value. Observations presented to the system essentially act as addresses, selecting the appropriate likelihoods to read from each array. The authors use stochastic computing techniques, generating random bit streams from the read likelihoods which get multiplied locally using AND gates. This architecture minimizes data movement. The prototype chip contained 2048 memristors and CMOS circuits which successfully performed inference. Simulations of a larger system showed orders of magnitude improvement in energy efficiency over a microcontroller implementation for a gesture recognition task. By mapping Bayesian networks directly into mixed-signal hardware using distributed memristor memory, this innovative work achieved significant acceleration and power savings compared to von Neumann architectures. It provides a blueprint for embedding sophisticated probabilistic reasoning on low-power edge devices.

At a high level, the Bayesian machine is implementing Bayes' theorem to compute posterior probabilities as shown in Eq. (5):

$$P(\text{Hypothesis}|\text{Data}) \propto P(\text{Data}|\text{Hypothesis}) \text{ x } P(\text{Hypothesis}) \tag{5}$$

More specifically in statistics language Eq. (5) translates into: Posterior $\propto$ Prior $\times$ Likelihood. Where likelihood is the prior probability of each hypothesis, stored in the "prior" memristor arrays. Prior is the likelihood of the observed data under each hypothesis, stored in the "likelihood" memristor arrays. The posterior is probability of each hypothesis given the data, computed by the Bayesian machine.

To perform this calculation:

1. The observed data acts as an address to look up the likelihood values from each likelihood array.

2. These likelihoods get converted to binary bit streams by comparison to random numbers.

3. The prior values are similarly converted to bit streams.

4. The bit streams are multiplied via AND gates, physically co-located with the arrays.

5. The resulting bit stream frequencies converge to the desired posterior probabilities.

This architecture maps the mathematical operations of Bayes' theorem into physical computing primitives using the memristor arrays for storage and localized analog/digital hardware for arithmetic. The stochastic computing handles the multiplication and summation while minimizing data movement. The distributed nature of the computation allows implementing sophisticated Bayesian models efficiently.

Harabi's prototype system demonstrated the potential for orders of magnitude improvements in energy efficiency compared to microprocessors. However, further optimizations in power could be made by using emerging devices like SOT-MRAMs for true random number generation rather than the LFSRs [53]. The binary architecture may also limit model accuracy compared to systems storing probabilities using multi-level memristors. While custom-designed for Bayesian processing, Harabi's digital approach could face challenges in matching the raw computational throughput of some analog in-memory neural network accelerators. Overall, the work makes excellent progress in tailoring memristor architectures to the specific needs of probabilistic inference, though analog and hybrid schemes may provide further advantages.

2.4 Memristor as MCMC sampler

While Harabi's work relied on stochastic computing, the implementation was rather complex and required complex electronics such as level shifters, and comparators. In a recent work Dalgaty et al. demonstrated a novel approach to utilize the intrinsic variability of memristive devices for machine learning [54]. Rather than viewing the cycle-to-cycle conductance fluctuations in devices like HfO_x Resistive Random Acces Memories (RRAMs) as a nuisance, the authors exploit it as a useful source of randomness. They configure a large RRAM array to physically realize a Bayesian machine learning model. The conductance states store the model parameters (likelihoods and priors) in a row-by-row fashion while the devices' stochastic switching implements Markov chain Monte Carlo (MCMC) sampling for training and inference based on the previous state.

Specifically, The $N \times M$ memristor array with total 16,384 memristors is used to store and compute the posterior distribution for a Bayesian model with M parameters. Each row n stores one sample model g_n, with each of the M differential memristor pairs in that row encoding one parameter of g_n.

The posterior is computed iteratively row-by-row using MCMC sampling. At each step, row n contains the current sample model g_n. To generate the next sample model g_{n+1} at row n + 1, the conductance of each memristor pair is updated by exploiting the stochastic switching of memristors. Applying a set voltage pulse samples a new conductance from a normal distribution centered at the current state. This randomly proposes a new model g_p that is either accepted or rejected.

If rejected, the same update step is repeated to propose a new model, with a digital counter C_n for row n incremented to track rejections. This counter represents the contribution of g_n to the posterior. After acceptance, the process advances to the next row. Over many rows, the distribution of stored conductance states converges to the target posterior distribution.

Despite memristor drift over 10 million switching cycles, the approach retains accuracy. Benchmarking indicates orders of magnitude efficiency gains over CMOS implementations of MCMC sampling with a typical energy consumption training being at 1 pJ per bit, with overall training expenditures being at 1–10 μJ whereas comparable task on consumer grade PC takes at most 1 J. By embracing rather than mitigating device non-idealities, this work opens a promising avenue for efficient in-situ Bayesian learning in memristive hardware.

2.5 Stochastic Bayesian computing with 2D materials

Previous examples show tremendous progress towards implementation of Bayesian machines. However, these approaches are based on conventional lithography techniques that frequently do not provide access to emerging materials and rapid prototyping techniques like printing for ultra-low power computing. Recently a materials science research group at Penn State University published two papers that descibes hardware implementation of Bayesian network using novel 2D memtransistors [55]. Similar to Harabi et al. [52] in this work, Bayesian networks in implementation was based on generating stochastic bits (s-bits) that randomly sample between 0 and 1 to represent probabilities.

The authors utilize monolayer molybdenum disulfide (MoS_2) memtransistors, where cycle-to-cycle fluctuations in the programmed conductance state provide inherent stochasticity. They design a compact s-bit generator circuit comprising just six memtransistors that can tune the probability of 1's in the bitstream from 0 to 1. This exploits the devices' non-volatile yet analog programmability.

Integrating multiple s-bit generators with 2D memtransistor-based logic gates, they demonstrate a hardware architecture for Bayesian network nodes. The approach is applied to a simple two-node network, accurately estimating conditional probabilities. By computing directly in memory, the architecture is estimated to be extremely energy efficient at around 1 pJ per s-bit.

The key innovation is developing fully in-memory Bayesian computing leveraging emerging devices. Unlike prior works using two-terminal memristors or spintronics that rely heavily on CMOS peripherals, the three-terminal 2D memtransistors enable standalone stochastic logic. This could unlock new applications in edge intelligence and probabilistic computing. The demonstration of area- and energy-efficient uncertainty handling highlights the potential of co-designing algorithms and nanodevices.

The same lab also presented another work where they utilized a combination of monolayer MoS_2 and WSe_2 memtransistors, leveraging their analog programmability and cycle-to-cycle conductance fluctuations. A compact circuit comprising just two MoS_2 memtransistors implements a stochastic synapse that can sample weights from configurable Gaussian distributions. Integrating MoS_2 and WSe_2 devices enabled modified tanh and sigmoid activation functions.

These synaptic and neuronal building blocks are then combined in a crossbar array architecture to physically implement BNN computations like a vector–matrix multiplication. The authors demonstrate a BNN circuit for disease classification using SPICE simulations. Uncertainty decomposition reveals the various entropy components.

Table 2 summarizes the differences of the works discussed in this chapter.

	Harabi et al.	**Zhang et al.**	**Dalgaty et al.**	**Sebastian et al.**
Architecture	Bayesian machine	BNN	BNN	BNN
Device technology	HfOx memristors	MoS2 memtransistors	HfOx memristors	MoS2, WSe2 memtransistors
Array size	16 × 8 × 8 (2048 devices)	6 memtransistors	16,384 devices	Not discussed
# of levels	Binary (LRS/HRS)	Analog	Binary (LRS/HRS)	Analog
Key mechanism	Binary conductance states and external CMOS stochastic blocks	Cycle-to-cycle variation for intrinsic stochasticity	Cycle-to-cycle variation for MCMC sampling	Analog conductance states and stochastic switching
Implementation	130 nm CMOS	Compact circuit demo	Computer-in-loop with memristor array	Crossbar BNN SPICE simulation
Training approach	In situ MCMC sampling	Pre-trained inference	In situ MCMC sampling	Pre-trained inference
Energy consumption	~2 pJ/bit	~2 pJ/bit	1 pJ/bit	18.37 nJ inference
Endurance	Not discussed	Not discussed	10 M cycles	Not discussed

Table 2.
Comparison of memristor/memtransistor based Bayesian computing approaches.

3. Conclusions

This work presented a range of toolkits for bio-inspired computing using memristors and neuristors, providing both original research and a literature review. In particular, we highlighted that memristors are versatile elements capable of storing information, performing computation in passive artificial neural networks, and being utilized in spiking neural networks and Bayesian machines. We demonstrated theoretically that passive memristors exhibit NDR when randomly connected in networks. Individual devices also show oscillatory properties, similar to an active neuristor, when cooling and heating at nanoscale are considered. Such oscillations could lead to spiking neural network implementations. In particular, a novel memristive device model exhibiting chaotic dynamics was introduced. The proposed model incorporates non-linear drift behavior based on the thermodynamic principles of Newtonian cooling and Joule heating within the memristive element. Numerical simulations demonstrate that this minimal model is capable of producing chaotic oscillations. The complex oscillatory characteristics and inherent stochasticity arising from the chaotic memristor suggest potential utility for hardware-based stochastic computing architectures. Further research is warranted to fully characterize the dynamical landscape and harness the rich nonlinear dynamics for unconventional computing paradigms.

Additionally, we explored an alternative approach of Bayesian computing with memristors. We reviewed some of the most recent works leveraging emerging memory technologies like memristors and memtransistors for efficient in-situ Bayesian inference. These approaches harness the intrinsic stochastic switching dynamics of such devices through either binary conductance fluctuations or analog variability. This avoids the need for expensive external CMOS random number generation and enables direct physical mapping of probabilistic models into memory arrays.

While these results are promising, memristor/memtransistor-based Bayesian computing remains in early development stages. Key near-term challenges to address include improving endurance, tolerance to variations, and energy efficiency. Integrating appropriate peripheral circuits for parallel readout and programming also needs to be solved before large-scale systems can be realized. Significant co-innovation across algorithms, circuit designs, and nanomaterials will be essential to unlock the full potential of in-memory Bayesian accelerators.

Longer-term research can explore hybrid CMOS-memristor architectures and systems integrating sensing, learning, and actuation for adaptive Bayesian edge intelligence. More complex Bayesian models like Bayesian neural networks and probabilistic deep learning equivalents should be mapped to further leverage emerging device properties. This co-design approach combining tailored algorithms, devices, and hardware for probabilistic inference may enable new ultra-low power intelligent systems for autonomous edge applications.

In conclusion, the intersection of Bayesian techniques and nanoscale memristive devices holds exciting promise to impart human-like uncertainty handling and adaptable intelligence to future electronics. Academic research leveraging their complementary strengths can establish the foundation for this next generation of probabilistic computing hardware.

Conflict of interest

The authors declare no conflict of interest.

Author details

Nikolay Frick
Materials Science and Engineering, North Carolina State University, Raleigh, USA

*Address all correspondence to: nvfrik@ncsu.edu

References

[1] Haldeman C, Beggs JM. Critical branching captures activity in living neural networks and maximizes the number of metastable states. Physical Review Letters. 2005;**94**(5). DOI: 10.1103/physrevlett.94.058101

[2] Fagerholm ED et al. Cortical entropy mutual information and scale-free dynamics in waking mice. Cerebral Cortex. 2016;**26**(10):3945-3952. DOI: 10.1093/cercor/bhw200

[3] Kinouchi O, Copelli M. Optimal dynamical range of excitable networks at criticality. Nature Physics. 2006;**2**(5): 348-351. DOI: 10.1038/nphys289

[4] Maass W. Networks of spiking neurons: The third generation of neural network models. Neural Networks. 1997; **10**(9):1659-1671. DOI: 10.1016/s0893-6080(97)00011-7

[5] Chistiakova M, Volgushev M. Heterosynaptic plasticity in the neocortex. Experimental Brain Research. 2009;**199**(3-4):377-390. DOI: 10.1007/s00221-009-1859-5

[6] Blomfield S. Arithmetical operations performed by nerve cells. Brain Research. 1974;**69**(1):115-124. DOI: 10.1016/0006-8993(74)90375-8

[7] Koch C. Biophysics of Computation. Oxford University Press; 1998. DOI: 10.1093/oso/9780195104912.001.0001

[8] Gerstner W et al. Neuronal Dynamics: From Single Neurons to Networks and Models of Cognition. Cambridge University Press; 2014. DOI: 10.1017/CBO9781107447615

[9] Krizhevsky A, Sutskever I, Hinton GE. Imagenet classification with deep convolutional neural networks. In: Advances in Neural Information Processing Systems. Vol 25. 2012

[10] BERT: Pre-Training of Deep Bidirectional Transformers for Language Understanding. Available from: https://arxiv.org/abs/1810.04805v2 [Accessed: September 21, 2023]

[11] Aaron van den Oord et al. WaveNet: A Generative Model for Raw Audio. 2016. DOI: 10.48550/ARXIV.1609.03499. Available from: https://arxiv.org/abs/1609.03499

[12] Sejnowski TJ. The unreasonable effectiveness of deep learning in artificial intelligence. Proceedings of the National Academy of Sciences. 2020;**117** (48):30033-30038. DOI: 10.1073/pnas.1907373117

[13] Esteva A et al. A guide to deep learning in healthcare. Nature Medicine. 2019;**25**(1):24-29. DOI: 10.1038/s41591-018-0316-z

[14] Gal Y, Ghahramani Z. Dropout as a Bayesian Approximation: Representing Model Uncertainty in Deep Learning. In: Balcan MF, Weinberger KQ, editors. Proceedings of The 33rd International Conference on Machine Learning. Vol. 48. New York, New York, USA: Proceedings of Machine Learning Research PMLR; 20-22 Jun 2016. pp. 1050-1059. Available from: https://proceedings.mlr.press/v48/gal16.html

[15] Srivastava N et al. Dropout: A Simple Way to Prevent Neural Networks from Overfitting. Journal of Machine Learning Research. 2014;**15**(56):1929-1958. Available from: http://jmlr.org/papers/v15/srivastava14a.html

[16] Radford A et al. Improving language understanding by generative

pre-training. 2018. Available from: https://gwern.net/doc/www/s3-us-west-2.amazonaws.com/d73fdc5ffa8627bce44dcda2fc012da638ffb158.pdf [Accessed: September 21, 2023]

[17] Chua L. Memristor-The missing circuit element. IEEE Transactions on Circuit Theory. 1971;**18**(5):507-519. DOI: 10.1109/tct.1971.1083337

[18] Strukov DB et al. The missing memristor found. Nature. 2008; **453**:80-83

[19] Frick N et al. Modeling and characterization of stochastic resistive switching in single Ag(2)S nanowires. Scientific Reports. 2022;**12**:6754

[20] Jo SH et al. Nanoscale memristor device as synapse in neuromorphic systems. Nano Letters. 2010;**10**(4): 1297-1301. DOI: 10.1021/nl904092h

[21] Joshua Yang J et al. Memristive switching mechanism for metal/oxide/metal nanodevices. Nature Nanotechnology. 2008;**3**(7):429-433. DOI: 10.1038/nnano.2008.160

[22] Du C et al. Reservoir computing using dynamic memristors for temporal information processing. Nature Communications. 2017;**8**(1). DOI: 10.1038/s41467-017-02337-y

[23] Kim KM et al. Electrically configurable electroforming and bipolar resistive switching in Pt/TiOsub2/sub/Pt structures. Nanotechnology. 2010;**21**(30):305203. DOI: 10.1088/0957-4484/21/30/305203

[24] Jeong DS et al. Emerging memories: Resistive switching mechanisms and current status. Reports on Progress in Physics. 2012;**75**(7): 076502. DOI: 10.1088/0034-4885/75/7/076502

[25] Jo SH, Kim K-H, Lu W. Programmable resistance switching in nanoscale two-terminal devices. Nano Letters. 2009;**9**(1):496-500. DOI: 10.1021/nl803669s

[26] Song SJ et al. Real-time identification of the evolution of conducting nano-filaments in TiO2 thin film ReRAM. Scientific Reports. 2013;**3**(1). DOI: 10.1038/srep03443

[27] Li Y et al. Ultrafast synaptic events in a chalcogenide memristor. Scientific Reports. 2013;**3**(1). DOI: 10.1038/srep01619

[28] Ohno T et al. Short-term plasticity and long-term potentiation mimicked in single inorganic synapses. Nature Materials. 2011;**10**(8):591-595. DOI: 10.1038/nmat3054

[29] Park S et al. RRAM-based synapse for neuromorphic system with pattern recognition function. In: 2012 International Electron Devices Meeting. IEEE; 2012. DOI: 10.1109/iedm.2012.6479016

[30] Crane HD. The Neuristor. IEEE Transactions on Electronic Computers. 1960;**EC-9**(3):370-371. DOI: 10.1109/tec.1960.5219861

[31] Crane H. Neuristor-A novel device and system concept. Proceedings of the IRE. 1962;**50**(10):2048-2060. DOI: 10.1109/jrproc.1962.288234

[32] Nagumo J, Arimoto S, Yoshizawa S. An active pulse transmission line simulating nerve axon. Proceedings of the IRE. 1962;**50**(10):2061-2070. DOI: 10.1109/jrproc.1962.288235

[33] Wilamowski BM, Czarnul Z, Bialko M. Novel inductorless neuristor line. Electronics Letters. 1975;**11**(15):355. DOI: 10.1049/el:19750270. DOI: 10.1049%2Fel%3A19750270

[34] Morin FJ. Oxides which show a metal-to-insulator transition at the Neel temperature. Physical Review Letters. 1959;**3**(1):34-36. DOI: 10.1103/physrevlett.3.34

[35] Pickett MD, Medeiros-Ribeiro G, Williams RS. A scalable neuristor built with Mott memristors. Nature Materials. 2012;**12**(2):114-117. DOI: 10.1038/nmat3510

[36] Kumar S, Strachan JP, Williams RS. Chaotic dynamics in nanoscale NbO2 Mott memristors for analogue computing. Nature. 2017;**548**(7667): 318-321. DOI: 10.1038/nature23307

[37] Driscoll T et al. Phase-transition driven memristive system. Applied Physics Letters. 2009;**95**(4). DOI: 10.1063/1.3187531

[38] Horrocks GA et al. Scalable hydrothermal synthesis of free-standing VOsub2/sub nanowires in the M1 phase. ACS Applied Materials & Interfaces. 2014;**6**(18):15726-15732. DOI: 10.1021/am504372t

[39] Li M et al. Hydrothermal synthesis of VOsub2/sub polymorphs: Advantages challenges and prospects for the application of energy efficient smart windows. Small. 2017;**13**(36). DOI: 10.1002/smll.201701147

[40] Pergament A et al. Electrical switching and oscillations in vanadium dioxide. Physica B: Condensed Matter. 2018;**536**:239-248. DOI: 10.1016/j.physb.2017.10.123

[41] Beaumont A et al. Current-induced electrical self-oscillations across out-of-plane threshold switches based on VO2 layers integrated in crossbars geometry. Journal of Applied Physics. Apr. 2014; **115**(15). DOI: 10.1063/1.4871543

[42] Crunteanu A et al. Voltage- and current-activated metal–insulator transition in VOsub2/sub-based electrical switches: A lifetime operation analysis. Science and Technology of Advanced Materials. 2010;**11**(6): 065002. DOI: 10.1088/1468-6996/11/6/065002

[43] del Valle J et al. A caloritronics-based Mott neuristor. Scientific Reports. 2020;**10**(1). DOI: 10.1038/s41598-020-61176-y

[44] Hodgkin AL, Huxley AF. A quantitative description of membrane current and its application to conduction and excitation in nerve. The Journal of Physiology. 1952;**117**(4):500-544. DOI: 10.1113/jphysiol.1952.sp004764

[45] Fangohr H et al. Joule heating in nanowires. Physical Review B. 2011;**84**(5). DOI: 10.1103/physrevb.84.054437

[46] Lashkare S et al. Transient Joule heating-based oscillator neuron for neuromorphic computing. IEEE Electron Device Letters. 2018;**39**(9):1437-1440. DOI: 10.1109/led.2018.2854732

[47] Liao Z-M et al. Evolution of resistive switching over bias duration of single Ag2S nanowires. Applied Physics Letters. 2010;**96**(20). DOI: 10.1063/1.3432665

[48] Camps O, Stavrinides SG, Picos R. Stochastic Computing Implementation of Chaotic Systems. Mathematics 2021;**9**(4). ISSN: 2227-7390. DOI: 10.3390/

math9040375. Available from: https://www.mdpi.com/2227-7390/9/4/375

[49] Torrejon J et al. Neuromorphic computing with nanoscale spintronic oscillators. Nature. 2017;**547**(7664):428-431. DOI: 10.1038/nature23011

[50] Indiveri G, Liu S-C. Memory and information processing in neuromorphic systems. Proceedings of the IEEE. 2015; **103**(8):1379-1397. DOI: 10.1109/jproc.2015.2444094

[51] Citri A, Malenka RC. Synaptic Plasticity: Multiple Forms Functions, and Mechanisms. Neuropsychopharmacology. Aug 2007;**33**(1):18-41. DOI: 10.1038/sj.npp.1301559

[52] Harabi K-E et al. A memristor-based Bayesian machine. Nature Electronics. Dec 2022. DOI: 10.1038/s41928-022-00886-9

[53] Fu Z et al. An Overview of Spintronic True Random Number Generator. Frontiers in Physics. Apr. 2021;**9**. DOI: 10.3389/fphy.2021.638207

[54] Dalgaty T et al. In situ learning using intrinsic memristor variability via Markov chain Monte Carlo sampling. Nature Electronics. Jan 2021;**4**(2): 151-161. DOI: 10.1038/s41928-020-00523-3

[55] Zheng Y et al. Hardware implementation of Bayesian network based on two-dimensional memtransistors. Nature Communications. Sept 2022;**13**(1). DOI: 10.1038/s41467-022-33053-x